Case Workbook in HUMAN GENETICS

Ricki Lewis

SUNY-Albany

WCB Wm. C. Brown Publishers
Dubuque, Iowa • Melbourne, Australia • Oxford, England

Book Team

Editor *Elizabeth M. Sievers*
Developmental Editor *Margaret J. Kemp*
Production Editor *Rachel K. Daack*
Art Editor *Joseph P. O'Connell*
Permissions Coordinator *Gail I. Wheatley*
Visuals/Design Developmental Consultant *Donna Slade*

Wm. C. Brown Publishers
A Division of Wm. C. Brown Communications, Inc.

Vice President and General Manager *Beverly Kolz*
Vice President, Publisher *Kevin Kane*
Vice President, Director of Sales and Marketing *Virginia S. Moffat*
National Sales Manager *Douglas J. DiNardo*
Marketing Manager *Patrick Reidy*
Advertising Manager *Janelle Keeffer*
Director of Production *Colleen A. Yonda*
Publishing Services Manager *Karen J. Slaght*
Permissions/Records Manager *Connie Allendorf*

Wm. C. Brown Communications, Inc.

President and Chief Executive Officer *G. Franklin Lewis*
Corporate Senior Vice President, President of WCB Manufacturing *Roger Meyer*
Corporate Senior Vice President and Chief Financial Officer *Robert Chesterman*

Cover photo © Grace Natoli-Sheldon/Third Coast Stock Source, Inc.

Copyright © 1994 by Wm. C. Brown Communications, Inc. All rights reserved

A Times Mirror Company

ISBN 0-697-22287-X

No part of this publication may be reproduced, stored in a retrieval system, or transmitted, in any form or by any means, electronic, mechanical, photocopying, recording, or otherwise, without the prior written permission of the publisher.

Printed in the United States of America by Wm. C. Brown Communications, Inc., 2460 Kerper Boulevard, Dubuque, IA 52001

10 9 8 7 6 5 4 3 2 1

TABLE OF CONTENTS

Acrocephalosyndactyly   1
Acute Leukemia   3
Adenosine Deaminase Deficiency   5
Alopecia Areata   7
Alstrom Syndrome   9
Amyotrophic Lateral Sclerosis   11
Anhidrotic Ectodermal Dysplasia   13
Argininemia   15
Bipolar Affective Disorder   17
Bloom Syndrome   19
Blue Diaper Syndrome   21
Breast Cancer   23
C2 Deficiency   25
Carnitine-acylcarnitine Translocase
   Deficiency   27
Carnosinemia   29
Chronic Granulomatous Disease   31
Cleft Lip/Cleft Palate   33
Congenital Contractural Arachnodactyly   35
Cystic Fibrosis   37
DiGeorge Syndrome   39
Enamel Hypoplasia   41
Epicanthus   43
Epidermolysis Bullosa   45
Facioscapulohumeral Muscular Dystrophy   47
Factor XI Deficiency   49
Familial Creutzfeldt-Jakob Disease   51
Familial Hypertrophic Cardiomyopathy   53
Familial Mediterranean Fever   55

Fatal Familial Insomnia   57
Fragile X Syndrome   59
Gonadal Dysgenesis   61
Gyrate Atrophy   63
Hearing Loss   65
Heart Attack   67
Hemochromatosis   69
Hemophilia A   71
Hyperkalemic Periodic Paralysis   73
Lattice Corneal Dystrophy   75
Leber Hereditary Optic Neuropathy   77
Li-Fraumeni Family Cancer Syndrome   79
Menkes Disease   81
Multiple Endocrine Neoplasia   83
Myotonic Dystrophy   85
Nephrolithiasis   87
Neurofibromatosis Type 2   89
Non-insulin Dependent Diabetes Mellitus   91
Osteopetrosis   93
Pelizaeus-Merzbacher Disease   95
Placental Aromatase Deficiency   97
Prader-Willi/Angelman Syndromes   99
Protein C Deficiency   101
Pseudohermaphroditism   103
Purine Nucleoside Phosphorylase
   Deficiency   105
Red-green Colorblindness   107
Schneckenbecken Dysplasia   109

Severe Childhood Autosomal Recessive
   Muscular Dystrophy 111
Sickle Cell Disease 113
Tangier Disease 115
Tay-Sachs Disease 117
Tourette Syndrome 119
Turner Syndrome 121
Vestibular Schwannoma 123
Von Hippel-Lindau Disease 125
Von Willebrand Disease 127

Appendix A 131

Human Genetics Glossary 133

A Glossary of Chemical Terms Relevant to Genetics 143

A Glossary of Disorders Mentioned in the Text 145

PREFACE TO CASES IN HUMAN GENETICS

The study of genetics is incomplete without applying the principles to real people in real situations. This short book is a collection of cases and studies culled from the recent medical literature. They build on concepts presented in your genetics or biology textbook and also introduce new information.

To get the most out of CASES IN HUMAN GENETICS, you will need to apply a little T.L.C.:

Tools and Technologies. The classic tools of the geneticist—the pedigree and the genetic code—are on the inside cover. Today, events at the cellular and molecular levels increasingly explain symptoms. From this multi-level view, comes a host of new technological tools that offer the great precision of diagnosing problems based ultimately on genotype, rather than, or to supplement, more traditional descriptions of phenotype. These new methodologies include RFLP genetic markers, Southern blotting, the polymerase chain reaction, and vastly improved cytogenetics based on DNA probes. The cases and studies ask you to apply these approaches explained in your text to dissect the genetic underpinnings of health and illness.

Logic. Applying Mendel's laws to predict recurrence risks and to identify carriers requires an understanding of probability, as well as a strong dose of logic. Would a person with a particular condition live long enough or feel well enough to have children? Why would an extremely rare disorder be unusually prevalent in a highly inbred and isolated population? Which treatments provide temporary relief from symptoms, and which offer more permanent solutions by altering genes? Which facts in a case are relevant to answering particular questions? Solving genetic problems goes beyond applying principles in a vacuum; it entails using those principles to better understand why and how a trait or illness develops.

Compassion. The narratives in this book often tell what it is like to live with a genetic disease. They go beyond the more familiar cystic fibrosis and hemophilia, sickle cell disease and color blindness. You probably have not heard of many of the disorders described in the following pages. I made an effort to offer examples not usually found in genetics textbooks, so you can apply concepts and skills to new situations. Included are a sampling of the extremely rare, 5000 or so inherited illnesses and traits, and also some fairly common conditions not often thought of as "genetic." For example, several entries deal with family cancer syndromes; and heart disease and diabetes are also related to altered gene activity.

More important than learning to recognize modes of inheritance or moving with ease from DNA to RNA to protein, is the appreciation of how inherited illness affects the quality of people's lives. When answering a question about a genetic marker test to presymptomatically diagnose a mental illness or cancer syndrome, imagine that you are the person facing the choice of learning how you will probably die. When two people find they are carriers of the same rare disorder by having a severely ill child, imagine the anxiety of undergoing a second pregnancy marked by various genetic tests. I hope that from these tales of real people, you learn not only genetics, but also compassion.

Organization

Each entry begins with a list of KEY WORDS to help you find the appropriate sections in a textbook. Next, a narrative focuses on people and their genes (with the exception of one entry on quarter horses). The stories usually concern a single extended family; they also include real investigations ranging from pilot studies on a few individuals, to large-scale efforts involving many hundreds of participants.

Questions follow the narrative. These include pedigree construction, predicting risks, identifying characteristics of genes and modes of inheritance, manipulating the genetic code, and drawing conclusions from research results.

Level of Difficulty

The cases and their questions vary in difficulty. If you are confused by the many mutations in the cystic fibrosis gene and what they mean, don't despair, so are geneticists. We also still do not understand the Angelman and Prader-Willi syndromes, different in phenotype, yet resulting from inheriting the apparently same two chunks of chromosome 15, with one disorder if it comes from the father, another if it comes from the mother. The expanding genes of myotonic dystrophy and fragile X syndrome explain some aspects of these disorders but also raise many new questions.

These confusing disorders illustrate that we geneticists can never grow complacent in our knowledge. Whenever we think that every fact is neatly shoehorned into an explanation, a paradigm of thought, another anomaly appears. It keeps us on our toes.

CASES IN HUMAN GENETICS will reinforce what you have learned in genetics or biology class and help you to remember the principles of genetics by placing them in the context of someone's life. These exercises test critical thinking skills, rather than recall, so they may help you in other subjects as well. Finally, I hope that these true cases will move some of you to consider a career in genetics; the number of genetic counselors and medical geneticists lags far behind the growing need.

Ricki Lewis

ACROCEPHALOSYNDACTYLY

Key Words

Mendelian inheritance
Pedigree analysis
Prenatal diagnosis

Wayne and Marge had always thought that their identical twin sons Willy and Todd had unusually large toes. This was cute when they were infants, but when they began to walk, their feet did not fit easily into shoes. The problem worsened as they grew older, but then they discovered a certain brand of running shoe that fit well. Wayne's mother Cecile recalled that she had had similar problems fitting Wayne and his sister Colleen with shoes when they were children. Cecile's deceased husband George had had very bizarre feet. Colleen and her husband Jack have a 4-year-old daughter Leah who has the family's big toes, too.

Colleen is a geneticist. She suspected that Willy, George, Todd, Wayne, Leah, and herself have a form of acrocephalosyndactyly, an inherited disorder. She suggested that the affected relatives have their toes x-rayed and, as she suspected, they all had the unusual sign of this condition—double bones in each toe!

Worksheet

1. Acrocephalosyndactyly in this family cannot be inherited as a sex-linked trait because _____ .

2. A prenatal diagnostic test for acrocephalosyndactyly might be _____ .

3. Complete the pedigree of this family if the big toe trait is inherited as an autosomal dominant condition.

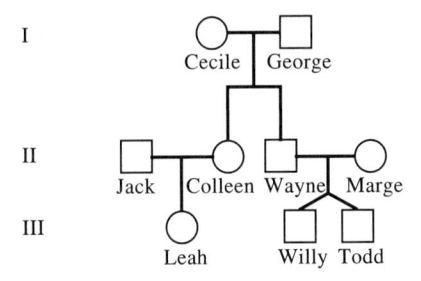

4. Complete the pedigree of this family if the big toe trait is inherited as an autosomal recessive condition, including individuals who must be carriers.

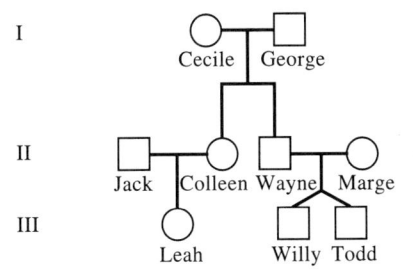

5. If the trait is inherited as an autosomal dominant, and Leah marries John, a ballet dancer with exquisite, slim toes, the probability that a child of theirs would also inherit normal toes is _____ .

ACUTE LEUKEMIA

Key Words

Alu sequences
Evolutionary conservation
Germline mutation
Homeotic mutant
Homology

Intron
Oncogene
Somatic mutation
Transcription factor
Tumor suppressor

People who develop acute (rapid onset and progression) leukemia often have a translocation that cuts chromosome 11 at a certain band in the long arm and places it on either chromosome 4 or 9. The aberration appears only in the leukemic cells. The translocation breakpoint on chromosome 11 occurs within an intron of a gene that encodes an 11.5 kilobase transcript. The breakpoint is within a cluster of sequences called "Alu." An Alu sequence is a 300 base pair segment repeated about 500,000 times throughout the human genome.

Leukemia arrests the normal process by which stem cells in the bone marrow give rise, through a series of developmental decisions, to several more specialized cell types. The acute leukemias associated with translocations affect mature white blood cells called monocytes and lymphocytes, which descend from the same stem cells.

The gene disturbed by the translocation in the human leukemias encodes a protein that is remarkably similar in sequence to a gene in the fruit fly *Drosophila melanogaster* called trithorax. Trithorax is a homeotic mutation, controlling segmentation and early embryogenesis. Wild type homeotic genes determine which developmental pathways a cell follows. When homeotic genes are mutant, cells develop into the "wrong" structure—such as a fly with legs where its antennae should be.

The human leukemia gene is very similar in sequence to genes in cows, pigs, rodents, and rabbits. In all of these species, the related genes encode a similar transcription factor.

Worksheet

1. How can two different translocations—t(4;11) and t(9:11)—cause the same phenotype?

2. What do leukemia in humans and homeotic mutations in fruit flies have in common conceptually that might explain the extensive homology between the human chromosome 11 leukemia gene and the fly's trithorax gene?

3. Explain how the chromosome 11 translocation might cause cancer if an oncogene is involved, compared to if a tumor suppressor gene is implicated.

4. How can a mutation in an intron, which is not represented in the gene's protein product, nevertheless affect the phenotype?

5. Why is acute leukemia not inherited?

6. How might the presence of highly repeated Alu sequences within an intron cause a translocation?

7. What is the evidence that the human leukemia gene may be quite ancient?

ADENOSINE DEAMINASE DEFICIENCY

Key Words

AIDS
B cells
Genotype
Germline gene therapy
Phenotype
Somatic gene therapy
T cells

A few dozen children in the world suffer from adenosine deaminase (ADA) deficiency. ADA is an enzyme necessary for the immune system's T cells to function. A type of T cell called a helper activates B cells, that then secrete antibody proteins that help in the immune defense against infection. Populations of T and B cells form the immune system. When ADA is deficient as a result of an autosomal recessive mutation, immunity is lacking.

Various ways to give these children the needed ADA are being tried:

- Chelsea received a bone marrow transplant from her father at 4 months of age. His healthy cells populated her bone marrow, replacing her defective cells with cells that produce the needed enzyme. She is doing well.
- Michael received unaltered ADA manufactured using recombinant DNA technology. Unfortunately, the enzyme stays in the blood only for a few minutes before being broken down—not long enough to restore immunity.
- Laura received ADA that was chemically linked to polyethylene glycol (PEG). This chemical allows ADA to remain in the bloodstream for a week. Weekly injections allowed Laura's immune system to develop to the point that she can live a nearly normal life. It still is not certain if she would survive a severe infection.
- Shelly was selected to undergo gene therapy when a bone marrow donor could not be found for her, PEG-ADA did not work, and her T cells took up an ADA gene delivered in a disabled retrovirus in culture. Mature T cells were separated from Shelly's blood, given the ADA gene, and reinfused into her body. Her immune system gradually began to function, but the treatment needed to be repeated every few weeks as the engineered T cells died.
- Todd had gene therapy like Shelly, but on more immature T cells taken from his bone marrow rather than from his blood. Todd soon had many corrected T cells in his blood, as well as antibodies for the very first time. He even grew tonsils, a part of the body absent in children with ADA deficiency. Todd's ADA activity rose to 25% of normal levels, enough to provide immunity. It is not yet known how long Todd can go until he needs another treatment, but it is far longer than Shelly can remain healthy between treatments.

Worksheet

1. State whether each child's treatment corrects the phenotype or the genotype.

2. Shelly needs to repeat her treatment more often than does Todd because _____.

3. Is the gene therapy performed on Shelly and Todd somatic or germline?

4. Shelly and Todd know each other from the medical center where they receive their gene therapy, and they have become good friends. If they grow up healthy and marry, the risk of their child inheriting ADA deficiency is _____ .

5. In acquired immune deficiency syndrome (AIDS), a virus attacks and kills helper T cells, leading to a decrease in B cell activity. AIDS, therefore, is a _____ of ADA deficiency.

ALOPECIA AREATA

Key Words

Pedigree
Phenotype
Mode of inheritance

Chrissy is a beautiful 17 year old who is very upset when her hair begins to fall out in big patches. Her sister Mandy, fearing she, too, will lose her tresses, asks their mother Pam if the condition is inherited. Pam confesses that she has worn a wig for years because she, too, has hair loss. The girls' maternal grandmother Liz and their father Joe are bald. Their maternal grandfather Adam, and paternal grandparents Charles and Polly all have full heads of hair. The condition in this family is called alopecia areata, and it is inherited as an autosomal dominant trait.

Worksheet

1. A phenocopy of alopecia areata is _____.

2. The sisters blame their hair loss on their bald father, because their mother has some hair. From which parent did they inherit alopecia? What is the evidence for this?

3. The chance that a child of Chrissy's will inherit alopecia is _____.

4. Draw a pedigree for this family.

ALSTROM SYNDROME

Key Words

Consanguinity   **Pedigree analysis**
Genetic markers   **Pleiotropy**
Mendelian inheritance

Paul and Diana have seven children. Two boys and two girls have four of the same medical problems—obesity, deafness, loss of central vision, and diabetes mellitus. Charles has a brother Ralph, married to Alice, and they have two daughters. Diana has a sister, Wendy, married to Joe, and they have three boys, Tom, Steve, and Russ. Paul's parents are Michael and Sandra; Diana's parents are Frank and Bertha. Frank is Michael's brother. Paul and Diana's family doctor thinks that the four children with the four symptoms—George, John, Becky, and Tabitha—have an autosomal recessive condition called Alstrom syndrome.

Worksheet

1. Draw the pedigree for this family, including all of the given information.

2. Further analysis of this family shows that Frank and Bertha are obese and hard of hearing; Wendy has diabetes and would easily be overweight if she didn't diet and exercise; and Paul has slight vision and hearing deficits. If these individuals also have Alstrom syndrome, then this particular gene is _____

   and _____ .

3. The steps in searching for a genetic marker for Alstrom syndrome in this family would be:

4. Alstrom syndrome is extremely rare. A fact about this family that might explain its appearance is that _____
   _____ .

5. If Paul and Diana have another child, the chance that he or she will inherit Alstrom syndrome is _____ .

AMYOTROPHIC LATERAL SCLEROSIS

Key Words

Genotype	Mutation
Mode of inheritance	Phenotype

Thomas and Sheldon meet in the waiting room of their neurologist, who is helping them each cope with the progressive muscle weakening of amyotrophic lateral sclerosis (ALS), which is also known as Lou Gehrig's disease. Both men have recently been diagnosed after experiencing slight stiffening and weakening of the legs and arms. Both are in their forties. Sheldon is very distressed over the diagnosis, because his mother and her father died from ALS within five years of diagnosis. He watched them gradually become quadriplegics and knows that they died as their respiratory muscles became paralyzed. Thomas has never heard of ALS. No one in his family has had it, and he can trace his family back five generations.

ALS affects 1 in 100,000 people worldwide, but only 10% of cases are inherited as an autosomal dominant trait. The other 90%, presumably environmental in origin, are termed sporadic. Researchers are beginning to understand the molecular underpinnings of the disorder.

ALS is caused by a mutation in a gene on chromosome 21 that encodes the enzyme superoxide dismutase (SOD1). This enzyme is an anti-oxidant, which means that it normally detoxifies free radicals, which are unstable by-products of metabolism that can damage tissue. SOD1 is a well-known substance—it is even sold in health food stores.

Worksheet

1. What further information is needed before SOD1 could be investigated as a treatment for ALS?

2. "At least you don't have to worry about passing ALS to your children, as I do," says Sheldon to Thomas. How might Sheldon be incorrect?

3. The risk that a child of Sheldon would inherit ALS is _____ .

ANHIDROTIC ECTODERMAL DYSPLASIA

Key Words

Cytogenetics
Evolutionary conservation
Gene mapping
Mode of inheritance
Translocation
X-inactivation

In 1875, Charles Darwin described an interesting family as follows:

> ... 10 men, in the course of 4 generations, were furnished, in both jaws taken together, with only 4 small and weak incisor teeth and with 8 posterior molars. The men thus affected have very little hair on the body, and become bald early in life. ... It is remarkable that no instance has occurred of a daughter being affected. ... Although the daughters are never affected, they transmit the tendency to their sons; and no case has occurred of a son transmitting it to his sons.

When additional families with the disorder, called anhidrotic ectodermal dysplasia, were studied, some females were found to be very mildly affected with small or malformed teeth and abnormal sweat glands. Their lack of hair and sweat glands appeared to be patchy, with some areas of skin affected and some not. A test to detect female carriers examines the distribution of sweat glands on the entire back. Carriers have patchy distributions.

The gene for anhidrotic ectodermal dysplasia is localized to chromosome Xq12. The map position was accomplished by studying a translocation between chromosomes X and 9 in a girl. A mutation called "tabby" in the mouse gives a similar phenotype and maps to the same region on the X chromosome.

Worksheet

1. The most likely mode of inheritance for this condition is _____.

2. Gene expression in females is "patchy" because of _____.

3. Is the gene for this disorder on the short or long arm of a chromosome?

4. The girl with the X;9 translocation might have children with medical problems because _____
   _____.

5. A female who shows some symptoms of a condition carried on the X chromosome is called a
   _____ _____ .

6. What is the evidence that the gene behind anhidrotic ectodermal dysplasia is ancient?

ARGININEMIA

Key Words

Enzyme
Exon
Genetic code
Genetic engineering

Inborn error of metabolism
Mode of inheritance
Multiple alleles
Mutation

Japanese researchers examined mutant alleles of a gene encoding an enzyme called liver-type arginase from four patients. The wild type protein catalyzes the breakdown of arginine, an amino acid. This inborn error of metabolism, argininemia, causes progressive mental retardation, spastic limb movements, seizures, and growth retardation.

A cDNA revealed that the coding portion of the gene specifies 322 amino acids. The entire gene is 11.5 kilobases and is localized to chromosome 6q. Argininemia affects both sexes and is inherited from two carrier parents.

Patient A was homozygous for a G mutated to an A at DNA base 365 in the liver-type arginase gene. Patient B was homozygous for a G to C mutation at base 703, which caused one amino acid to be replaced by another. Patient C had patient A's mutation and patient B's mutation. Patient D had patient A's mutation in one allele, and the other allele was a deletion of a C at position 842.

Each type of mutation was functionally evaluated by cloning it in *E. coli*. Patient A's mutant protein was too short. The other mutations yielded proteins of normal length that were unstable or otherwise nonfunctional.

Worksheet

1. The mode of inheritance of argininemia is _____.

2. The patients who are heterozygotes for the argininemia gene are _____ and _____.

3. The patients who have missense mutations are _____ and _____.

4. Patient A's liver-type arginase is too short because_____.

5. Human liver-type arginase can be synthesized and expressed in *E. coli* because
    a. *E. coli* have livers.
    b. the genetic code is universal.
    c. the disorder is recessive.
    d. *E. coli* also uses arginine.
    e. the genetic code is triplet.

6. The argininemia gene has enough bases beyond those in exons to encode _____ more amino acids.

BIPOLAR AFFECTIVE DISORDER

Key Words

Linkage
Mode of inheritance
Multifactorial inheritance
Pedigree
Penetrance

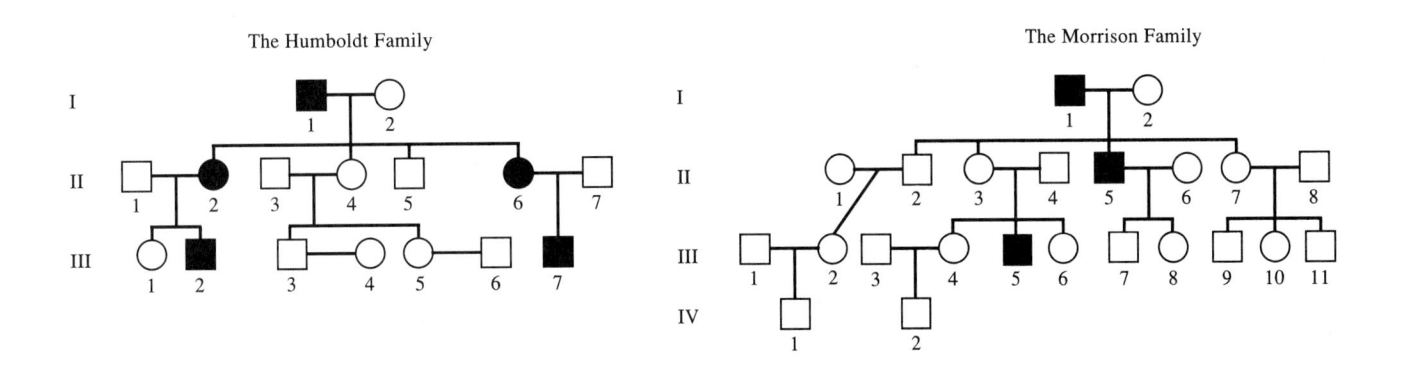

The Humboldt and Morrison families each have two medical conditions—red-green colorblindness and bipolar affective disorder (also known as manic-depression), a mental illness characterized by extreme mood swings. Colorblindness is sex-linked recessive. Bipolar affective disorder can be inherited as a sex-linked recessive trait or as an autosomal recessive trait.

In the Humboldt family, individuals I1, II2, II6, III2, and III7 suffer from episodes of bipolar affective disorder and are colorblind. These individuals are also the only ones in the family to have a certain RFLP on the X chromosome.

In the Morrison family, the symbols for individuals with bipolar affective disorder are completely shaded. The colorblind individuals are II2, III5, IV1, and IV2. All of the people with bipolar affective disorder also have an informative RFLP on chromosome 11, as do individuals II3 and II7.

Worksheet

1. In the Humboldt family, the mode of inheritance for bipolar affective disorder is _____ _____.

   Carriers of the disorder are individuals _____ and _____.

2. In the Morrison family, the mode of inheritance for bipolar affective disorder is _____ _____.

   Carriers of the disorder are individuals _____, _____, _____, and _____.

3. What is the evidence for the modes of inheritance of bipolar affective disorder in each family?

4. What factors might complicate tracing the inheritance pattern of bipolar affective disorder?

5. If Humboldt III2 marries Morrison III8, the chance that a son of theirs has bipolar affective disorder is _____.

BLOOM SYNDROME

Key Words

Cancer
Consanguinity
DNA repair
DNA replication
Environment

Founder effect
Gene therapy
Hardy Weinberg equilibrium
Ligase
Phenotype

Only 100 people in the world have the autosomal recessive Bloom syndrome. Twenty-four of these people have parents who were cousins, and many of the affected families come from isolated communities. The disorder is caused by a mutation in the gene encoding ligase I, an enzyme that connects the sugar-phosphate backbone in newly replicated DNA.

At the cellular level, signs of Bloom syndrome include:

- slow growth of cells in culture (slowed cell cycle)
- increased frequency of chromosome breakage and rearrangements
- increased sister chromatid exchange
- DNA replication forks that progress very slowly
- a delay in the polymerization of Okazaki fragments into larger DNA molecules
- increased sensitivity to DNA damaging agents

E. coli and yeast cells with mutant ligases display similar characteristics when grown in culture.

Whole body symptoms of Bloom syndrome include:

- increased risk of cancer, particularly leukemia
- facial rash due to sensitivity to the sun
- severe immune impairment

The oldest Bloom syndrome patient lived into middle age, but most die much younger.

Worksheet

1. How does the cellular phenotype explain the whole body phenotype?

2. The symptom that indicates a defect in DNA repair is _____.

3. Gene therapy to treat Bloom syndrome would be difficult because _____.

4. Evidence that the frequency of the mutant allele causing Bloom syndrome is not in Hardy Weinberg

   equilibrium is that _____.

5. How does the environment affect expression of Bloom syndrome?

6. How does the mechanism of carcinogenesis in Bloom syndrome differ from that of a cancer caused by an oncogene, and from that of a cancer caused by a mutant tumor suppressor gene?

BLUE DIAPER SYNDROME

Key Words

Mode of inheritance
Pedigree

Gloria adores her nephews William and Charlie. The boys have an unusual, but harmless, inherited condition. Because of abnormal transport of the amino acid tryptophan across the small intestinal lining, bacteria act on urine precursors to produce a compound that turns indigo blue upon contact with the air. William's parents were quite alarmed just after his birth when his wet diapers turned blue! By the time they had Charlie, they were accustomed to the blue urine.

The family doctor assured Gloria and the boys' parents, Gloria's sister Edith and her husband Archie, that other family members needn't worry about the condition recurring. He explained that it is autosomal recessive and appeared in William and Charlie because Edith and Archie are both carriers. The condition had not occurred in past generations and likely wouldn't again unless Edith and Archie had more children, each of whom stood a 25% chance of inheriting the trait.

Gloria and her husband Michael are, therefore, surprised when their fraternal twins Marshall and Joey each produce blue diapers!

Worksheet

1. The error that William and Charlie's physician made was that _____.

2. Draw a pedigree for this extended family and their unusual disorder.

3. Gloria and Edith wonder if their grandsons will inherit blue diaper syndrome. Will they? Why or why not?

4. The blue diaper stain is likely to be deeper after a person eats food containing a lot of _____.

BREAST CANCER

Key Words

Cancer	**Mode of inheritance**
Conditional probability	**Oncogene**
Genetic heterogeneity	**Somatic mutation**
Germline mutation	**Tumor suppressor**

Breast cancer affects 1 in 10 women at some time in their lives. Of these women, 5% have inherited a germline mutation in a gene called BRCA1 on chromosome 17q, which provides a susceptibility. Tumors develop when a somatic mutation also affects the other BRCA1 gene on the homologous chromosome in breast cells. In the other 95% of breast cancer patients, two somatic mutations in the BRCA1 gene cause the disorder. The germline form of breast cancer tends to occur in younger women in both breasts and raises the risk of cancer developing elsewhere, particularly in the ovaries. Mutations in BRCA1 cause about 70% of germline breast cancers.

Breast cancer can be successfully treated with surgery, chemotherapy, and radiation if it is detected early. Women with an affected relative are urged to have mammograms (a special breast X ray) every few years to detect the condition as early as possible.

Joyce is 32 years old. Her mother had breast cancer at age 37, and now Joyce's sister Winifred has just found a tiny lump in her breast that is cancerous. Winifred warns Joyce that she has a 50% risk of inheriting the gene that causes breast cancer, and, therefore, she has a 50% risk of developing breast cancer.

Worksheet

1. Is Joyce's risk exactly 50%, or might it be higher or lower? Cite a reason for your answer.

2. How can studying women with inherited breast cancer supply knowledge that could help women whose breast cancer is not caused by a germline mutation?

3. Of all women, what is the chance that an individual will develop familial (germline) breast cancer?

4. What is the evidence that inherited breast cancer is genetically heterogeneic?

5. A man can develop breast cancer due to a mutation in the androgen receptor gene on the X chromosome. The cancer apparently develops from a relative overexpression of estrogen, rather than from activation of a cancer-causing gene. These facts suggest that breast cancer is

    a. sex limited and genetically heterogeneic.
    b. sex linked recessive and not genetically heterogeneic.
    c. caused by an oncogene and a tumor suppressor.
    d. not sex limited, and it is genetically heterogeneic.

C2 DEFICIENCY

Key Words

Control of gene expression **Multifactorial inheritance**
Mode of inheritance **Pedigree analysis**

The C2 gene encodes the C2 protein, which is part of an immune system biochemical pathway called the complement system. The C2 gene resides in the HLA (human leukocyte antigen) gene complex on chromosome 6.

A person with two mutant C2 alleles has impaired immunity and suffers frequent bacterial infections. Carriers can be detected by determining HLA type. Two types of defects in C2 gene function are known. In 94% of affected families, no C2 protein is detectable. This is type I C2 deficiency. In 6% of families, C2 protein is synthesized but cannot be secreted from cells. This is type II.

C2 deficiency can be difficult to diagnose because many conditions can impair immunity against bacterial infection. In the Fonebone family, the frequent severe infections in two brothers led to the diagnosis. Jerome Fonebone was hospitalized six times in three years for strep, cellulitis, croup, and various unexplained fevers. His brother Antone had repeated bouts of pneumonia and sinusitis. A sister, Mona, is very healthy, as is her mother Joan. Their father Tyrone, however, suffers often from bronchitis, sinusitis, and pneumonia. HLA typing revealed that Joan is a carrier for C2 deficiency.

Worksheet

1. The mode of inheritance for C2 deficiency is _____ _____ .

2. Draw a pedigree for the Fonebone family.

3. Is Mona a carrier of C2 deficiency?

4. In what sense is C2 deficiency a multifactorial trait?

5. How could you distinguish between types I and II C2 deficiency at the molecular level?

6. How could C2 deficiency be non-penetrant?

CARNITINE-ACYLCARNITINE TRANSLOCASE DEFICIENCY

Key Words

Gene therapy
Mitochondrial inheritance
Modes of inheritance

Jim D. died at 4 days of age, two days after suffering cardiac arrest. The pregnancy had been uneventful, and he had seemed normal at birth. His older sister was healthy. Two years after Jim's death, his parents had another son.

Like Jim, Kerry seemed normal at birth. But at 36 hours, his heart rate plummeted; he had a seizure; and he stopped breathing. He was successfully resuscitated. The doctor, knowing Jim's illness, ordered several tests and found excess long-chain fatty acids in the blood. These are components of triglyceride fats that are broken down by mitochondria in the liver to provide glucose for energy during starvation. It was, therefore, significant that Kerry had not eaten since birth.

Kerry was prescribed a special infant formula low in fatty acids, and his parents were instructed to feed him often. Despite this treatment, the child was frequently hospitalized for vomiting, lethargy, an enlarged liver, and poor muscle control, and once for becoming comatose after a feeding was delayed. It would take two to three weeks to recover from each episode. His weakness worsened, and triglycerides began to accumulate in his muscles. His liver enlarged. Finally, just before his third birthday, Kerry died in the hospital of respiratory failure.

Kerry and Jim had a rare inborn error of metabolism called carnitine-acylcarnitine translocase deficiency. This is an autosomal recessive condition in which an enzyme produced in the mitochondria is lacking.

Worksheet

1. The probability that Kerry and Jim's sister is a carrier of this disorder is _____.

2. The children's mother is taking a biology course, and she learns that mitochondria are passed from the mother only. When the doctor explains how her sons had abnormal mitochondria, she feels guilty for having passed the disorder on to them. What is the evidence that she is incorrect in believing her mitochondrial gene made the boys ill?

3. An organ that would be a good target for somatic gene therapy for this condition is _____.

CARNOSINEMIA

Key Words

Amino acids
Mendelian inheritance
PKU

Simon and Adrienne Jackson are healthy people who lived in a rural area with poor health services when they had their first child Benjie. He had seizures from infancy, and as he grew into toddlerhood, it became apparent that he was severely mentally retarded. He died at 26 months of age. Because he had never been diagnosed with a specific disorder, an autopsy was performed. His brain showed signs of great derangement, with nerve cells degenerating and missing. No diagnosis was made, and because no other relatives had been affected, a genetic problem was not suspected. The family physician assured the couple that the condition was not likely to repeat.

The Jacksons waited a few years, then had another child shortly after moving to Chicago. Sadly, little Julie had the same symptoms as her brother did. This time, the parents took her to a major medical center where urine and cerebrospinal fluid tests revealed large amounts of a chemical called carnosine. This is a dipeptide consisting of the amino acids alanine and histidine joined together. A medical geneticist told of this finding tested the urine of the parents. Each had half the normal activity for an enzyme called carnosinase. Julie has and Benjie had an inherited disorder, carnosinemia.

Worksheet

1. The mode of inheritance for carnosinemia in this family is _____.

2. How can the parents have half the normal activity for this enzyme yet be healthy?

3. What type of chemical bond joins alanine to histidine to form carnosine?

4. How might you devise a dietary treatment for carnosinemia, similar to that for PKU?

5. In one experiment on two children with carnosinemia, all sources of dietary protein with an alanine next to a histidine were eliminated from the diet. The children still excreted carnosine in the urine. An explanation for

    this finding might be that _____
    _____.

CHRONIC GRANULOMATOUS DISEASE

Key Words

Environment
Genetic heterogeneity
Intron
Mutation

Pedigree analysis
Prenatal diagnosis
Sex-linked inheritance
X inactivation

A 67-year-old man is hospitalized with fever, chills, headache, and lack of appetite due to infection by the bacterium *Pseudomonas cepacia*. Not only is this particular infection very rare, but also it killed the man's 5-year-old grandson four years earlier. The grandson had been diagnosed with chronic granulomatous disease, a sex-linked recessive condition in which certain white blood cells can ingest microbes but cannot then destroy them as white blood cells normally would. The grandfather apparently had chronic granulomatous disease, too, although, oddly, he had been healthy until the present infection. Most people with this disorder have frequent bacterial infections.

Healthy white blood cells called phagocytes kill engulfed bacteria by producing a chemical called superoxide, which is formed when oxygen picks up electrons. The electrons are passed by a complex of four proteins, each of which is encoded by a different gene. Chronic granulomatous disease results from an abnormality in one of these proteins whose gene, called gp91-phox (phox stands for phagocyte oxidase), resides on the X chromosome.

The gp91-phox gene is sequenced from the grandfather, and the gene product is extracted from his phagocytes and is sequenced. Ten amino acids are missing from the carboxyl end of the protein, a severe enough abnormality to impair its function. The corresponding gene, however, has a single nucleotide base mutation—an A is changed to a G in a region of the gene that does not encode protein but is found between regions that do encode protein.

When the family is told that the same underlying illness that killed the grandson is causing the grandfather's infection, other family members are tested, specifically the females, to see if they are carriers. This is easy to do—a blue dye called tetrazolium will change color if superoxide is present. The grandfather's daughter and her two daughters (by her second husband) have some cells that have superoxide and some that do not.

The pedigree for the family is:

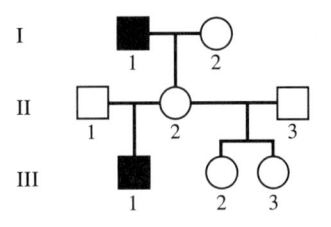

Worksheet

1. The grandfather's daughter (individual II2) is pregnant. She is very worried that a son would be ill, as her first son was, but her present husband (individual II3) assures her that is not possible because he is the father now. Is the husband correct?

2. Was the tetrazolium test for carrier status necessary for the grandfather's daughter?

3. In what part of the gene is the mutation in this family?

4. How can a single base change lead to a loss of 10 amino acids?

5. Is the mutation causing chronic granulomatous disease in this family a transition or a transversion?

6. White blood cells are obtained from the fetus, and they lack superoxide activity. Further genetic information that might be useful in evaluating the fetus' future is _____.

7. How does the environment influence the expression of the gp91-phox gene?

8. How might chronic granulomatous disease be inherited as an autosomal recessive trait?

CLEFT LIP WITH OR WITHOUT CLEFT PALATE

Key Words

Degree of relationship **Multifactorial inheritance**
Heritability **Pedigree analysis**

Cleft lip (failure of the upper lip to fuse) and cleft palate (failure of the soft palate to fuse) are multifactorial anomalies that occur during prenatal development. More than 200 syndromes include cleft lip with or without cleft palate. When clefts occur alone, heritability is estimated to be from 77% to 97%. Individuals with an affected relative have an increased risk of being born with a cleft. The risk of recurrence depends upon the degree of relationship to an affected individual as follows:

- if one parent has a cleft, risk to the child is 4%
- if one sibling has a cleft, risk to other sibling is 4%
- if two siblings have a cleft, risk to other sibling is 9%
- if one parent and one sibling have a cleft, the risk to another sibling is 17%

Worksheet

1. In the following families, a couple expecting a child is concerned that she or he will have a cleft lip with or without cleft palate. For each couple, cite the risk that the fetus (designated ◇) faces.

   a.
   b.
   c.
   d.
   e.

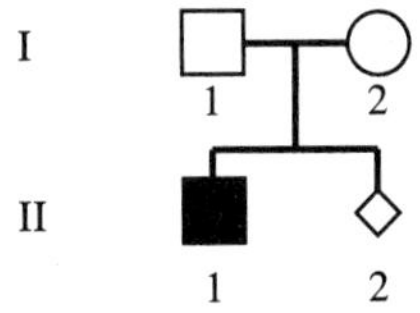

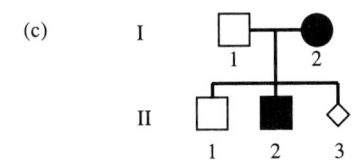

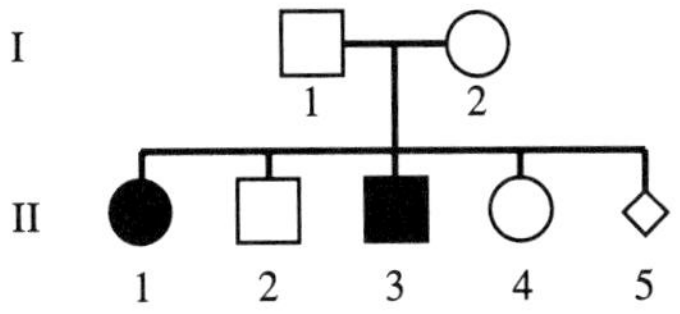

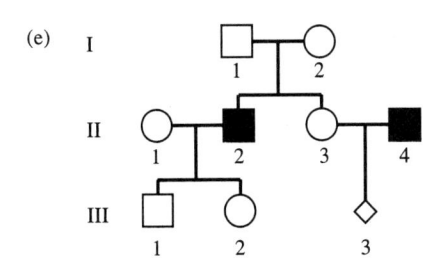

2. An estimate of the likelihood of recurrence based on population statistics for similar individuals is called the

   _____ risk.

3. Heritability means _____ .

CONGENITAL CONTRACTURAL ARACHNODACTYLY

Key Words

Genetic heterogeneity
Genotype
Linkage
LOD score
Mode of inheritance
Phenotype

The Throndson family consists of Rob and Natasha and their three children Jess, Morris, and Sue. Rob is very tall, with very long arms and legs. Morris and Sue have body builds similar to their father. Rob, Morris, and Sue also have weak muscles and unusual "crinkly" looking ears. Jess and Natasha are of shorter stature, with strong muscles and normal ears.

At a party a guest who is a medical geneticist sees the family together, notes the unusual features of Rob, Morris, and Sue, and invites them to meet with him at a nearby medical center. The geneticist initially suspects that the father and children have Marfan syndrome, but they do not have mitral valve prolapse, a defect in the heart valves often seen in Marfan's. Normal heart valves and crinkly ears are, however, consistent with congenital contractural arachnodactyly, an illness similar to Marfan syndrome but less likely to involve the deadly rupture of the aorta that often claims young Marfan patients.

Both Marfan syndrome and congenital contractural arachnodactyly are caused by mutations in genes for fibrillin, a large glycoprotein that forms microfibrils with other connective tissue proteins, particularly elastin. The genes behind the two autosomal dominant disorders, however, map to different chromosomes.

Specifically, genetic marker studies yield an LOD score of 6.2 for linkage of congenital contractural arachnodactyly to chromosome 5 and a negative LOD score when tested with markers to chromosome 15, the locus of the Marfan fibrillin gene. The Marfan phenotype segregates with markers to chromosome 15 with an LOD score of 25.6 but not with markers to chromosome 5, which yields a negative LOD score.

Worksheet

1. Construct a pedigree depicting the Throndson family.

2. The risk that a child of Natasha and Rob would inherit congenital contractural arachnodactyly is _____.

3. An LOD score of 6.2 means that
   a. the congenital contractural arachnodactyly gene is 6.2 kilobases long.
   b. the coding region of the congenital contractural arachnodactyly gene is 6.2 kilobases long.
   c. a person is 6.2 times as likely to develop congenital contractural arachnodactyly if he or she has an affected parent than if no other family members are affected.
   d. the odds of linkage between the congenital contractural arachnodactyly gene and markers on chromosome 5 are $10^{6.2}$ to 1.
   e. the genes for congenital contractural arachnodactyly and Marfan syndrome are linked.

4. Genetic heterogeneity is expected for the Marfan and congenital contractural arachnodactyly phenotypes

   because _____.

5. The fact that congenital contractural arachnodactyly affects a variety of tissues, reflecting the distribution of its fibrillin gene product, means that the gene is

   a. incompletely penetrant.
   b. multifactorial.
   c. pleiotropic.
   d. variably expressive.
   e. dominant.

CYSTIC FIBROSIS

Key Words

Carrier screening
Cystic fibrosis
Genetic heterogeneity
Genetic testing
Mendelian inheritance

Multiple alleles
Penetrance
Pleiotropy
Prenatal diagnosis
Reproductive technologies

Bryant is a 23-year-old graduate student in microbiology. He has cystic fibrosis, although he didn't know this until he was 8 years old. Before that, doctors simply treated him for infection after infection—usually bronchitis or pneumonia. He was terrified of catching a cold, which could land him in the hospital. Bryant also had abdominal pains a few hours after eating, and as a result, he was always severely underweight. Doctors noted a "sensitive stomach" in his chart, lectured his mother on providing proper nutrition, and attributed Bryant's pains to a nervous disposition.

A brother 2 years younger than Bryant named Will was very healthy; but when a sister Katie was born when Bryant was 8, a pattern appeared. Katie, like Bryant, battled very frequent respiratory infections. When as a baby she would scream after eating, doctors at first diagnosed colic and advised switching infant formulas. Nothing helped. When, at two years of age, Katie weighed only 20 pounds, her pediatrician began to put together her symptoms with those of her elder brother. A sweat test identified an abnormality in chloride transport characteristic of cystic fibrosis. Bryant and Katie were the only family members with the disorder.

Worksheet

1. In the following pedigree, fill in half of the symbols of individuals who must be carriers. Put a question mark in the symbols of those individuals who could be carriers.

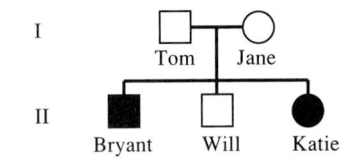

2. If Jane and Tom, the parents of these three children, decide to have another child, the chance that he or she will have cystic fibrosis is _____ ; that he or she will be a carrier for the condition is _____ ; the chance that a healthy child is a carrier is _____ .

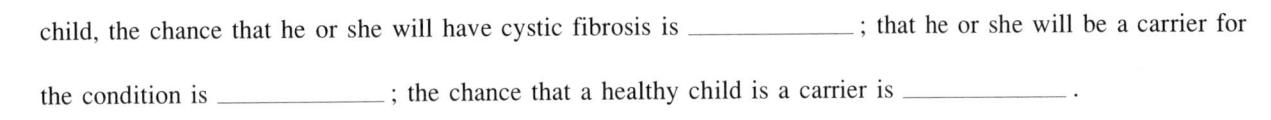

3. When Bryant decides to have children, assuming he is fertile, what type of test might his wife take to find out if she carries the cystic fibrosis gene?

4. The fact that cystic fibrosis produces symptoms in several organ systems illustrates a characteristic of a gene called

    a. genetic heterogeneity.
    b. pleiotropy.
    c. penetrance.
    d. multiple alleles.
    e. linkage.

5. If Jane and Tom decide to have another child, but want to ensure that he or she does not have cystic fibrosis, five tests and/or procedures that they might undergo are:

DIGEORGE SYNDROME

Key Words

Meiosis
Translocation

The P1 and F1 generations of the Findley family shown here are healthy, so they are surprised when Laura and her husband Aaron, and Dylan and his wife Iris have children. Karen, Leslie, and Lance have a rare condition called DiGeorge syndrome. The children have small facial features and low, rotated ears. More serious symptoms are defects in the blood vessels leading from the heart, an underdeveloped thyroid gland (impairing immunity), and underdeveloped parathyroid glands (disrupting calcium metabolism).

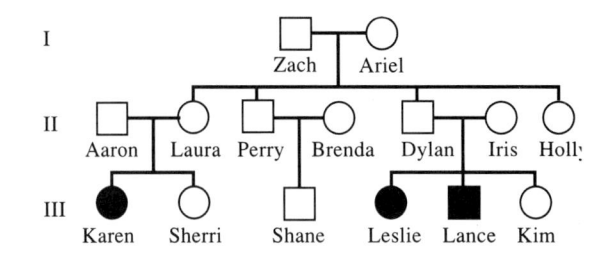

Dylan and Iris see a genetic counselor, Janet, who is concerned that their niece Karen has the same syndrome as Leslie and Lance. Janet learns that Iris has had several spontaneous abortions, as has her sister-in-law Laura. Noting the combination of repeated birth defects and pregnancy losses, Janet suggests that Dylan and Laura, their siblings, children, and parents be karyotyped. The results are, unfortunately, as Janet suspects.

Perry, Holly, Ariel, Sherri, and Shane have normal chromosomes. Laura, Dylan, Zach, and Kim have a reciprocal translocation between chromosomes 20 and 22. They have no symptoms but are translocation carriers with the following chromosomes:

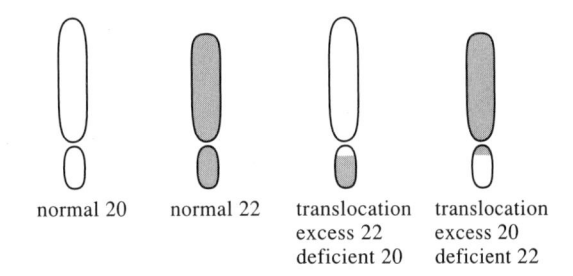

normal 20   normal 22   translocation   translocation
                          excess 22   excess 20
                          deficient 20   deficient 22

The children with DiGeorge syndrome, Karen, Leslie, and Lance, have partial monosomy 22 and partial trisomy 20, a genetic imbalance responsible for their symptoms. The reverse situation—partial trisomy chromosome 22 and partial monosomy 20—probably accounts for Laura and Iris' spontaneous abortions because these imbalances are incompatible with life.

Worksheet

1. Zach is an only child. His mother had three spontaneous abortions before giving birth to him, then she had a baby that died shortly after birth, which Zach barely remembers. Diagram the distribution of chromosomes 20 and 22 in meiosis and fertilization to depict how these pregnancy problems arose.

2. Why did Iris have reproductive difficulties if she is not a blood relative of Zach and has normal chromosomes?

3. Karen, Leslie, and Lance have
    a. two normal copies of chromosomes 20 and 22.
    b. two normal chromosome 20s, one normal chromosome 22, one chromosome 22 with some chromosome 20 material replacing some chromosome 22 material.
    c. Two normal chromosome 22s, one normal chromosome 20, one chromosome 20 with some chromosome 22 material replacing some chromosome 20 material.
    d. One normal chromosome 20, one normal chromosome 22, and two translocated chromosomes.

4. To be a translocation carrier like Zach, Laura, Dylan, and Kim, one must inherit _____ of the translocated chromosomes.
    a. one
    b. both
    c. neither

ENAMEL HYPOPLASIA

Key Words

Incomplete penetrance **Pedigree analysis**
Mendelian inheritance **Variable expressivity**

Enamel hypoplasia is an hereditary defect of the baby teeth in which holes and cracks appear around the crowns of the teeth. It is inherited as an autosomal dominant trait, but with incomplete penetrance and variable expressivity. Below is a pedigree showing the Barker family, which has this trait, and the Needlemeyer family, with whom they marry.

Worksheet

1. From the pedigree, it appears that the individual who displays incomplete penetrance for this trait is _____ .

2. How might the variable expressivity appear among the affected members of the Barker family?

```
I      ● — □         ○ — □
      Ruth Langley Barker  Julie  Paul

II        ■ ————————— ○
         Chuck           Tara

III  ○ — □   ■ — ○       □ — ○
    Hazel Mark Larry    Casey Selma

IV      ■              ○      ○
      Herbert         Irene  Shirley
```

3. Irene, as a baby, liked to go to sleep with a bottle of juice. The sugar washing over her teeth caused terrible decay, so the distinctive marks of enamel hypoplasia could not be seen. A type of molecular test, using information from several family members, that might tell Irene if she has inherited enamel hypoplasia is a

   _____ test.

4. The phenomenon of an environmental health problem, such as tooth decay caused by going to sleep with a juice bottle in the mouth, resembling a known inherited disorder is called a _____ .

5. The probability that Herbert's child will inherit enamel hypoplasia is _____ .

6. The Barkers clean their attic and find old baby photos that shed some light on the family tooth anomaly. Ruth's mother Lucy had a toddler's grin full of odd shaped teeth. Her father Jerry had normal teeth. Ruth also discovers that her older brother Fred and older sister Anna had the abnormal teeth, but she remembers that her younger sister Lulubelle had beautiful, healthy teeth. Add this information to the pedigree.

EPICANTHUS

Key Words

Concordance
Mendelian inheritance
Penetrance

Epicanthus is folding of the upper eyelid. It is inherited as an autosomal dominant trait and is completely penetrant. Tom and Jim are the twins of Donald and Tina. Donald has epicanthus, but Tina does not.

Worksheet

1. If Tom and Jim are identical twins, the chance that both have epicanthus is _____.

2. If Tom and Jim are fraternal twins, the chance that both have epicanthus is _____.

3. The concordance value for this trait is _____.

4. If Donald and Tina have another child, the chance that he or she will have normal eyes is _____.

EPIDERMOLYSIS BULLOSA

Key Words

Gene therapy
Genetic heterogeneity
Modes of inheritance
Pedigree analysis

Epidermolysis bullosa (EB) is a disorder in which the skin blisters very easily, sometimes leaving permanent scars. There are several inherited forms and also an acquired, autoimmune form of the illness in which autoantibodies attack proteins in the skin.

Parents whose children have EB are chatting while waiting to see their dermatologist. The Renfrews have a son and a daughter, Bryan and Cheri, who each have the dystrophic type of EB. This is caused by a recessive mutation in a gene encoding type VII collagen. This is a connective tissue protein that forms fibers anchoring the lower skin layer, the dermis, to the upper layer, the epidermis. It is a very disfiguring form of the illness. Bryan and Cheri, who are twins, have numerous blisters on their skin, which will leave scars. Rita Renfrew is pregnant, and she and her husband Ronald are frightened that their next child will be affected too.

The Blackwells also have two children who have EB, Chad and Jeremy aged 5 and 7. They have a milder, "simplex" form of the illness, with blisters that form on their hands and feet during warm weather but do not leave scars. Their mother Beulah also has the condition, which is inherited as an autosomal dominant trait. EB simplex is caused by a mutation in a gene on chromosome 5 encoding keratin, a protein abundant in the epidermis.

Another couple without children sits in a corner. The Starkey's son Richard died in infancy of the severe "basement membrane" form of EB. He had inherited a recessive allele from each parent that made him unable to produce a protein called epiligrin that anchors the epidermis to a layer called the basement membrane.

Worksheet

1. Explain the genetic heterogeneity of EB. That is, how can similar phenotypes have different genotypes?

2. If Cheri Renfrew marries Jeremy Blackwell, the probability that their child inherits EB simplex is _____.

3. Draw a pedigree for the Renfrew family.

4. Gene therapy may be possible for Chad and Jeremy, because keratinocytes (skin cells filled with keratin) can grow in culture, accept foreign genes, and be grafted onto a person. Why wouldn't this approach help Bryan and Cheri or future children of the Starkey's who inherit EB?

FACIOSCAPULOHUMERAL MUSCULAR DYSTROPHY

Key Words

Candidate gene
Gene mapping
Genetic markers
Homeobox genes
Mendelian inheritance
Phenotype
Translocation (cytogenetics)

Members of the Quacken family have facioscapulohumeral muscular dystrophy (FSHMD). One of the least debilitating of the muscular dystrophies, this autosomal dominant condition causes weak facial muscles, shoulders, and upper arms, usually on one side of the body. Symptoms usually begin in adolescence. Harry Quacken is 68 years old and has the condition, as does his son Marvin and Marvin's son Kevin. Marvin's sister Joan is unaffected. Cytogenetic analysis shows that Harry, Marvin, and Kevin each have a balanced translocation that exchanges the tip of one copy of chromosome 4 with the tip of chromosome 11.

Worksheet

1. Marvin and his wife Mamie have, in addition to Kevin, a healthy daughter Alison. They have also had some less successful pregnancies—three miscarriages and a stillborn boy with several anomalies. An explanation for these reproductive problems, given the family history, might be _____.

2. Alison watches the Jerry Lewis Muscular Dystrophy telethon and becomes very upset after seeing wheelchair-bound boys with Duchenne muscular dystrophy. How does facioscapulohumeral muscular dystrophy differ from Duchenne muscular dystrophy in phenotype and genotype?

3. A genetic marker is identified localizing the gene for FSHMD to the tip of chromosome 4. It is difficult, however, to obtain flanking markers for this gene because _____
_____
_____.

4. Researchers studying homeobox genes find that a clone of such a gene hybridizes near the marker found for FSHMD. Three lines of evidence that might implicate the homeobox gene as a candidate gene for FSHMD are:

    1.

    2.

    3.

5. How might the phenotype associated with homeotic genes explain the phenotype of FSHMD?

FACTOR XI DEFICIENCY

Key Words

Exons
Gene therapy
Introns
Mutation

Population genetics
RFLP analysis
Southern blotting

In Israel the blood clotting disorder factor XI deficiency is fairly common, affecting 1 in 190 people, compared to one in a million in other populations. The phenotype ranges from no symptoms at all to excessive bleeding with injuries, surgery, or childbirth. The disorder is inherited as an autosomal recessive trait. The gene is 23 kilobases long and consists of 15 exons and 14 introns.

At a blood disorders clinic in Israel, three young people meet who are seeking medical help because they bleed greatly during dental procedures. Although each is diagnosed with factor XI deficiency, each has a different mutation in the factor XI gene:

- Joshua has a point mutation at an intron/exon boundary. The gene is transcribed into an mRNA specifying 18 extra amino acids.
- Jacob has a CUU mRNA codon at amino acid position 117 altered to an AUU.
- Isaac has a CAA mRNA codon at amino acid position 283 altered to a UAA.

The mutations can be detected because each either deletes or adds a restriction enzyme cutting site.

Worksheet

1. The name of the patient who makes too short a factor XI protein is _____.

    We know this because _____.

2. The name of the patient with a missense mutation is _____.

3. Cite two hypotheses to explain why factor XI deficiency is so much more common among the Ashkenazi population in Israel than elsewhere.

4. Suggest a somatic gene therapy for factor XI deficiency.

5. A gene that is 23 kilobases long could encode about 7666 amino acids. The factor XI protein is considerably smaller than this because _____.

6. Would deletion of a restriction enzyme cutting site produce DNA pieces larger or smaller than normal?

FAMILIAL CREUTZFELDT-JAKOB DISEASE

Key Words

Conditional probability	**Penetrance**
Genotype	**Polymerase chain reaction**
Mode of inheritance	**RFLPs**
Mutation	**Wild type**

Creutzfeldt-Jakob disease (CJD) is a rare degeneration of the brain, causing headache, progressive dementia, seizures, and death within a year of symptom onset. Most cases are sporadic; 5 to 15% of cases are inherited as an autosomal dominant trait with 0.56 penetrance.

CJD is one of a few disorders believed to be caused by an infectious particle consisting only of protein, called a prion. Biologists disagree over whether a prion is an agent, like a virus, or a protein made in response to infection. Irrespective of this ongoing debate, a gene on chromosome 20p, the prion related protein gene (PRNP), appears to be implicated in development of certain disorders. The wild type PRNP gene encodes the protein portion of a glycoprotein that accumulates in fibers outside cells in a few disorders including CJD.

In a large family that emigrated from Germany to the U.S. at the turn of the last century, nine individuals died of CJD, all between the ages of 45 and 70. DNA from two members who died while the family was being studied at the National Institute of Neurological Disorders and Stroke revealed a mutation in the PRNP gene. GAG is mutated to AAG at codon 200.

The location of the mutation within the PRNP gene was fortuitous (for the researchers) because it eliminated a recognition site for the restriction enzyme BSMA1. This means that adding BSMA1 to DNA from a family member with a mutant PRNP allele will produce characteristic large fragments.

To test this hypothesis, DNA from the two deceased family members and from nine close relatives was amplified. PCR primers were used that correspond to opposite ends of the coding region of the PRNP gene. PCR was performed for thirty-five cycles, and the resulting DNA fragments were separated and displayed using agarose gel electrophoresis. The results were as follows:

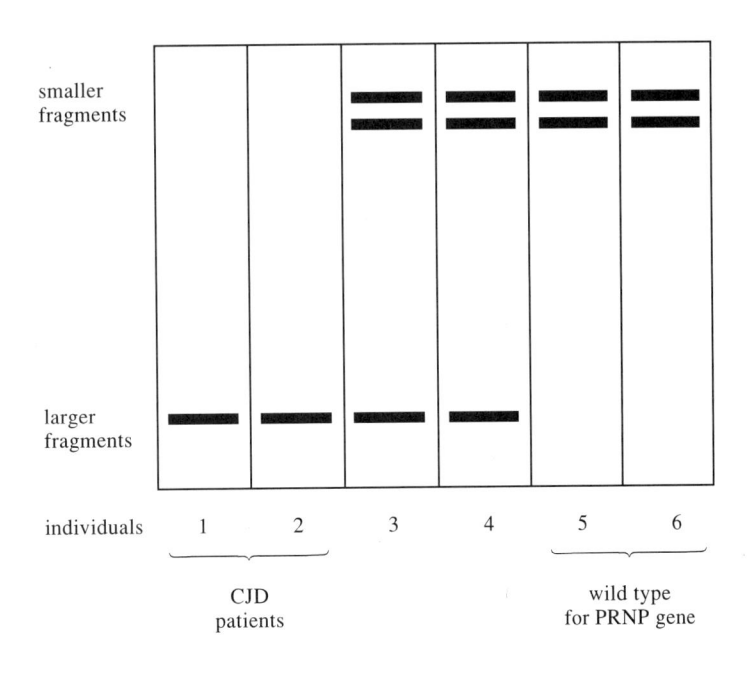

Worksheet

1. The mutation in codon 200 of the PRNP gene in this family alters the protein product by _____ .

2. This mutation is
   a. missense.
   b. nonsense.
   c. frameshift.
   d. at the chromosomal level.
   e. a translocation.

3. In the restriction fragment analysis of the mutant PRNP gene, the number of copies of the gene following amplification was
   a. 35.
   b. 70.
   c. 35^2.
   d. 2^{35}.
   e. 1,000,000.

4. What is the genotype for the PRNP gene for individuals 3 and 4 in the electrophoresis diagram?

5. An 18 year old in this family whose father died of CJD but whose mother is homozygous wild type wants to know his risk of also developing the condition. It is _____ .

FAMILIAL HYPERTROPHIC CARDIOMYOPATHY

Key Words

Genetic code
Genetic heterogeneity
Mutation

One cause of sudden death from heart failure in young adults is familial hypertrophic cardiomyopathy (FHC). The autosomal dominant condition is caused by a mutation in the gene encoding a part of the multi-subunit muscle protein myosin, called the heavy chain. Different mutations are found in different affected families. Severity of the illness correlates with the site of the mutation.

In the Winchester family, 34-year-old Paul died of heart failure while playing basketball. Because he had been healthy, an autopsy was performed. The nature of his heart muscle overgrowth, combined with a family history of his mother, a maternal aunt, and his maternal grandfather succumbing to heart failure in their thirties, led to a geneticist's request to analyze the cardiac myosin heavy chain gene. The geneticist found that Paul and a younger brother had a mutation at amino acid position 606, which changes a valine (val) to a methionine (met).

In the Churchill and Esposito families, different mutations cause FHC. These are more severe, causing death in the later teens and early twenties. Among the Churchills, arginine (arg) mutates to glutamine (gln) at position 249. In the Esposito family, an arginine mutates to a cysteine (cys).

Worksheet

1. A myosin molecule is shaped like a rod with a bulge at one end, called the head. The head is vital to muscle function. It is here that actin (the other major muscle protein) binds, that ATP splits to provide the energy needed for muscle contraction, and that the myosin light chain binds. A reason why mutations are not seen in the part of the gene encoding the myosin head is that _____

   _____ .

2. Cite two ways that the Churchill's FHC mutation can occur.

3. Cite three ways that the Esposito's FHC mutation can occur.

4. Why are there more members of the Winchester family affected with FHC than there are in the Churchill or Esposito families?

5. Would you expect FHC to be genetically heterogeneic? Why or why not?

6. Is the mutation in the Winchester family a transition or a transversion?

FAMILIAL MEDITERRANEAN FEVER

Key Words

Candidate genes **Gene mapping**
Consanguinity   **Pedigree analysis**
Cytogenetics   **Population genetics**

Familial Mediterranean fever, also known as periodic fever, is an autosomal recessive condition prevalent in certain population groups. Sufferers endure attacks of fever and abdominal and joint pain that last one to two weeks. Many white blood cells collect at the affected areas causing great inflammation. Death comes from kidney failure, and on autopsy the kidneys are blocked up with a gummy protein called amyloid. Daily oral doses of the drug colchicine can control symptoms and extend life.

One study examined DNA from twenty-seven Israeli families with affected members. The researchers screened more than 100 genetic markers (RFLPs), representing nearly all of the human chromosomes, against the DNA from the families. They found two markers, each mapping to the short arm of chromosome 16.

Worksheet

1. Another type of evidence that would confirm the localization of the familial Mediterranean fever gene to

   chromosome 16 would be _____

   _____ .

2. The high incidence of familial Mediterranean fever in certain ethnic groups can be attributed to the founder effect and subsequent consanguinious marriages or balanced polymorphism. What information is needed to distinguish between these two explanations?

3. An Armenian family living in Los Angeles has the pedigree to the right for familial Mediterranean fever. Indicate known carriers with a half filled-in symbol and possible carriers with a question mark.

4. What is the relationship between individuals III 1 and III 2 in the pedigree?

   a. brother and sister
   b. uncle and niece
   c. aunt and nephew
   d. first cousin and first cousin

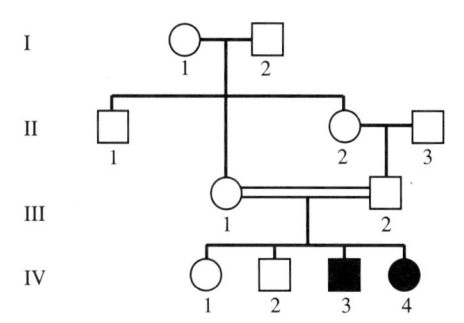

5. Familial Mediterranean fever is prevalent in natives of Turkey, Armenia, Italy, in North African Jews, and in middle eastern Arabs. The marker on chromosome 16 was found in Jews of North African ancestry living in Israel. If studies on other ethnic groups point to genes on different chromosomes, this would illustrate the concept of _____ .

6. Near the markers on chromosome 16 for familial Mediterranean fever are a cluster of genes that encode cell surface receptors to which white blood cells bind. This might explain the inflammation seen in the disorder. What experiments could be performed to test whether these genes are "candidate genes" for familial Mediterranean fever?

FATAL FAMILIAL INSOMNIA

Key Words

Gene therapy   **Pleiotropic**
Genetic code   **Presymptomatic diagnosis**
Mode of inheritance

A family in northern Italy has a horrifying, inherited illness in which degeneration of the sleep center of the brain makes some parts of sleep impossible—slow wave sleep and rapid eye movement (R.E.M.) sleep. Records from the 19th century allow the mode of inheritance to be traced. Of 288 relatives over six generations, twenty-nine people have had the disorder. There are no carriers; both sexes are affected; and every generation is affected until, by chance, no one inherits the mutant gene.

Symptoms begin at the average age of 49 years and include inability to produce tears, skin blotches, inability to feel pain, and poor reflexes, as well as the disturbed sleep. The lack of normal sleep eventually causes emotional instability, hallucinations, stupor, coma, and finally death thirteen months after symptom onset. Body functions that follow circadian rhythms, such as blood pressure and the levels of some hormones, become abnormal as the sleep-wake cycle is disrupted.

Fatal familial insomnia is caused by a point mutation in codon 178 of the gene for a protein called prion protein.

Worksheet

1. The people in this family in generation five are still in their childbearing years and do not want to pass the disorder onto their children. This is difficult to avoid because the age of onset is later than childbearing age. A technology that might be able to tell young people in this family whether or not they have inherited the

   causative gene is _____ .

2. Assuming that the gene behind this disorder has no introns, how many amino acids into the protein is the mutation located?

3. What information is needed to suggest gene therapy to treat fatal familial insomnia?

4. The mode of inheritance of fatal familial insomnia is _____ _____ .

5. Fatal familial insomnia illustrates pleiotropy because _____ .

FRAGILE X SYNDROME

Key Words

Anticipation
CpG island
Cytogenetics
Fragile X syndrome
Meiosis
Methylation
Mutation
Myotonic dystrophy
Pedigree analysis
Penetrance
Polymorphism
Sex-linked inheritance

Darcy was learning disabled with an IQ at the lowest end of normal. She finished high school, married, and started a family. Her first son Sly was healthy.

When Sly was two years old, Darcy and David had Barry. He was developmentally delayed, not reaching milestones such as sitting and standing when others his age would. He did not respond to stimulation as a typically curious infant or toddler would. Odd symptoms included a long face with coarse features, lopsided large ears, and large testicles.

At Barry's one year check-up, the pediatrician asked if any family members were mentally retarded or other children developmentally delayed. Darcy mentioned her sister Marcy's 10-month-old son Joseph who was not yet sitting.

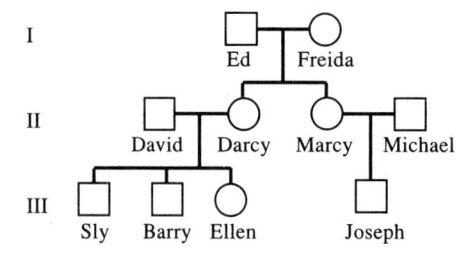

The physician told them that he thought Barry might have fragile X syndrome. This most common form of inherited mental retardation, affecting 1 in 1250 males and 1 in 2000 females, can be detected at three levels: whole body symptoms, abnormal chromosomes, and DNA level mutation.

Barry had all of the whole body symptoms—mental retardation, characteristic face, and large testicles, as well as the distinctive cytogenetics. When chromosomes from a severely affected fragile X patient are cultured in medium lacking the vitamin folic acid, a characteristic constriction can be seen near the end of chromosome Xq. A second cytogenetic sign is excess methylation (addition of CH_3 groups) of a CpG island, which is a stretch of GC-rich sequence preceding the structural gene called FMR-1 (for fragile X mental retardation-1).

In individuals wild type for fragile X syndrome, a trinucleotide, CGG, repeats five to fifty-four times just before the FMR-1 gene. In persons with severe fragile X syndrome, like Barry, this sequence repeats hundreds of times.

The gene causing fragile X and its surrounding areas of interest is depicted below:

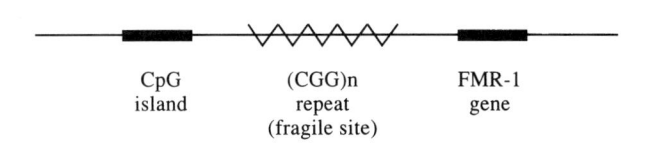

CpG island   (CGG)n repeat (fragile site)   FMR-1 gene

Fragile X syndrome is one of a few inherited conditions (see myotonic dystrophy) that displays "anticipation"—worsening of symptoms with each generation. Often a severely affected boy will have a mildly retarded grandfather. In many fragile X families, anticipation has a physical basis in the increasing CGG repeats. A grandfather might have an original unstable X chromosome that he passes to his daughter. The daughter has a 30% chance of suffering mild symptoms, such as mental slowness. Carrier females have more than the wild type number of CGG repeats but not as many as their fully affected sons do. The gene actually grows from generation to generation.

The excess methylation of the CpG island in severely affected boys is not seen in their carrier mothers. The fragile X phenotype is incompletely penetrant. Twenty percent of males who have the same microscopic symptoms as carrier females (intermediate increase in CGG number and normal methylation of the CpG island) have no symptoms. Nonpenetrant males and carrier females are considered to have a premutation for fragile X syndrome.

Darcy's extended family undergoes cytogenetic and DNA tests, with the following results:

Family Member	Fragile X Chromosome	CGG Repeats
Sly	No	25
Barry	Yes	960
Ellen	Yes	85
Joseph	Yes	900
Darcy	Yes	80
Marcy	Yes	82
Ed	Yes	94
Freida	No	18

Worksheet

1. Complete the pedigree, indicating carriers who have the fragile X premutation and those affected by the syndrome.

2. Suggest a mechanism to explain the growth of the CGG region preceding the FMR-1 gene with each generation.

3. Incomplete penetrance in female carriers can be explained by _____ .

4. A son of Ellen's faces a _____ probability of inheriting fragile X syndrome. He is likely to be severely affected because _____ .

5. A medical journal described a 24-year-old man with classic fragile X syndrome symptoms who was unusual because his X chromosome lacks a fragile site, the number of CGG repeats falls within the normal range, and his CpG island is normally methylated. The amino acid sequence, however, encoded by his FMR-1 gene has asparagine (asn) substituted for isoleucine (ile) at position 367. This finding suggests that

   a. the fragile X phenotype can result from events near the gene, and from a mutation within the gene.
   b. the man does not have fragile X syndrome, but Down syndrome or some other cause of mental retardation.
   c. the point mutation in the FMR-1 gene is a polymorphism unrelated to the man's phenotype.
   d. an error was made in the cytogenetics laboratory.

GONADAL DYSGENESIS

Key Words

Cytogenetics
Polymerase chain reaction
Sex determination
Transcription factor

Translocation
Turner syndrome
Y chromosome

Stella Walsh won a gold medal in the 100 meter dash in the 1932 Olympics. In 1980 she was killed in a robbery. When a physician examining the body noticed that she had ambiguous genitalia, she ordered a chromosome test. Stella had an XY karyotype. The same was true for another athlete, Ewa Klobukowska, who set a world record for the 100 meter dash in 1965. She, too, turned out to be XY.

Stella and Ewa had a condition called gonadal dysgenesis. They had the partial form of the condition, which meant that their genitals and gonads (testes and ovaries) were a mixture of male and female structures although all cells were XY. In complete gonadal dysgenesis, the phenotype is completely female on the outside but with "streak" gonads—just fibers of ovarian tissue where the ovaries should be.

Had the athletes competed in 1968 or later, they would have discovered their unusual conditions because in that year the International Olympic Committee began requiring a "buccal smear test" in which cheek cells are examined to determine if the person is XX or XY. The chromosome constitution was the final word on gender for competition purposes.

But it turned out that there is more to sex determination than the presence of a Y chromosome. Specifically, a single gene on the Y, the sex determining region gene or SRY, determines maleness—even if that gene is translocated to another chromosome. In 1982 researchers at the Whitehead Institute in Cambridge identified the region of the Y chromosome containing the SRY gene by studying two interesting groups of people:

- men with an XX karyotype and an SRY-containing a piece of the Y translocated to one X chromosome.
- women whose karyotype is XY, but their Y chromosome lacks the SRY gene.

In 1990 British researchers identified the SRY gene. It is a transcription factor that steers development towards maleness.

Worksheet

1. What must have been unusual about the Y chromosomes of Stella Walsh and Ewa Klobukowska?

2. In 1992 the International Olympic Committee replaced the buccal smear test with the polymerase chain reaction (PCR) to identify the SRY gene. How is this a more accurate assessment of gender than a karyotype?

3. What is unusual about the sex chromosomes of a human male who is XO?

4. How does the SRY gene direct development towards maleness?

GYRATE ATROPHY

Key Words

Exon **Maternal inheritance**
Genetic code **Mutation**

Often we do not know how a biochemical anomaly causes whole body symptoms. This is the case for gyrate atrophy, a degeneration of the retina that begins in late adolescence as night blindness and progresses to blindness. The cause is a mutation in the gene encoding a mitochondrial enzyme, ornithine aminotransferase (OAT), on chromosome 10. This autosomal recessive inborn error of metabolism causes the compound ornithine, a derivative of the amino acid arginine in dietary protein, to build up in body fluids. The only resulting symptom is the eye degeneration.

The OAT gene was sequenced for five patients with gyrate atrophy with the following results:

patient A: A change in codon 209 of UAU to UAA.
patient B: A change in codon 299 of UAC to UAG.
patient C: A change in codon 426 of CGA to UGA.
patient D: A 2 base deletion at codons 64 and 65 results in a UGA codon at position 79.
patient E: Exon 6, consisting of 1,071 bases, is entirely deleted.

Worksheet

1. Patient _____ has both a frameshift and a nonsense mutation.

2. Patients A, B, and C have in common _____ _____ .

3. Another patient, F, has the mutations seen in patient A and in patient B. How is this possible?

4. Gyrate atrophy does not exhibit maternal (mitochondrial) inheritance because _____ .

5. Patient E is missing _____ amino acids.

6. Suggest a way to relieve or slow the symptoms of gyrate atrophy.

HEARING LOSS

Key Words

Mendelian inheritance
Mitochondrial inheritance
Mode of inheritance

There are more than 100 inherited forms of hearing loss and many other non-inherited causes. Four young people with impaired hearing meet at a clinic, each deaf by a different mechanism. Explain how they differ in terms of mode of inheritance.

1. George has diabetes as well as hearing loss, due to a mutation in a mitochondrial gene.
2. Elroy is an albino in addition to his hearing impairment, which he inherited from his mother who is a carrier. His father is unaffected.
3. Jane has a form of osteogenesis imperfecta inherited on chromosome 17, which produces bone and connective tissue deformities and hearing loss. Her brother and father have the condition too.
4. Judy has Usher syndrome, whose symptoms include hearing loss and loss of central vision. Her sister has the condition, but neither of her parents do.

Worksheet

1.

2.

3.

4.

HEART ATTACK

Key Words

Environment
Genotype

Hardy Weinberg equilibrium
Multifactorial traits

Physicians have long been puzzled by people who suffer heart attacks but do not have any of the established risk factors of obesity, high serum cholesterol, smoking, fatty diet, or lack of exercise. Perhaps these people are genetically predisposed to develop heart disease. One candidate gene for heart attack susceptibility encodes angiotensin converting enzyme (ACE). Drugs that inhibit ACE open up blood vessels, improving heart function. Mutations in the ACE gene may, therefore, disrupt heart function, raising risk of a heart attack.

The ACE gene has two alleles designated D and I. The rarer allele, D, encodes a shortened enzyme that is the variant associated with elevated risk of heart attack. A team of French researchers recorded the three possible genotypes for this gene in populations in Belfast and three French cities. They found that over the three generations studied, the two alleles are in Hardy Weinberg equilibrium. The people differed in ACE genotype but not in other heart disease risk factors. One genotype, DD, is indeed associated with increased incidence of heart attack.

Worksheet

1. The other two genotypes of the ACE gene are _____ and _____ .

2. Why would evaluating ACE genotype as a predictor of heart attack risk be more accurate than using obesity, serum cholesterol, or fatty diet?

3. The observation that the two ACE alleles are in Hardy Weinberg equilibrium means that

   a. evolution is taking place for this gene.
   b. the allele frequencies are not changing from generation to generation.
   c. nonrandom mating is occurring.
   d. there is a great deal of consanguinity (inbreeding).
   e. the D allele and I allele have the same frequency.

4. If the D allele predisposes to heart attack, why does it persist? Cite two possible explanations for the gene's high frequency.

HEMOCHROMATOSIS

Key Words

Mode of inheritance
Penetrance
Sex influenced inheritance
Sex limited inheritance
Sex linked inheritance

In hemochromatosis the small intestine absorbs too much iron from food. Inheritance of two copies of a mutant gene on chromosome 6 causes the condition. Iron is deposited in the liver, heart, pancreas, endocrine glands, and skin, damaging these organs and causing hormone abnormalities and bronze skin. About 80% of individuals with hemochromatosis symptoms are men because women do not develop symptoms until after menopause (cessation of menstrual periods). The condition is treated by periodically removing blood. This must start before the heart and liver are damaged. The hemochromatosis gene is located within the HLA gene complex.

In the Carson family, brothers Dean and Jerry find out that they have hemochromatosis when they both develop liver cirrhosis and a bronze skin tone when they are in their forties. Their younger sister Carol has no symptoms, but comparisons of her HLA genes to those of her brothers reveal that she, too, has inherited the condition.

Worksheet

1. The mode of inheritance of hemochromatosis is _____.

2. Hemochromatosis is a _____ trait.

    a. sex limited
    b. sex influenced
    c. sex linked

3. If Carol's husband Henry is a carrier for hemochromatosis, the chance that a child of theirs inherits the disorder is _____.

4. Many people who have hemochromatosis do not realize that they have the condition because they do not eat iron-rich foods, so symptoms do not develop. This means that the condition is

    a. incompletely dominant.
    b. incompletely penetrant.
    c. incompletely expressive.
    d. variably recessive.

HEMOPHILIA A

Key Words

Conditional probability **Mode of inheritance**
Genetic marker **Pedigree**

Joanna is 28 years old and is thinking about starting a family. She is concerned because her brother and a cousin on her mother's side have hemophilia A.

In the 1980s Joanna had become convinced that when she became pregnant, she would have prenatal diagnosis so that she could discontinue the pregnancy if the fetus had inherited hemophilia. She had based this decision on the anxiety she felt that her brother might contract AIDS through the factor VIII gene product clotting factor that he often received. According to the National Hemophilia Foundation, a person with severe hemophilia might encounter the blood of 100,000 people in just one year by receiving donated clotting factor.

In 1993, however, a form of factor VIII produced with recombinant DNA technology became available, making treatment much safer. The new clotting factor can, also, be given preventively, greatly improving the quality of life for those with hemophilia. Joanna's conviction to avoid having an affected child began to change. But she still wants to know whether or not she is a carrier.

Joanna would like a more definitive estimate of her risk of being a carrier than can be derived from Mendel's laws. The amount of factor VIII in her blood is measured and found to be 20% of normal. To further confirm her carrier status, Joanna has her X chromosome checked to see if she inherited the same X that her brother did. Genetic markers confirm that this is indeed the case.

Worksheet

1. Draw a pedigree depicting the relationship of Joanna to her brother and cousin.

2. The chance that Joanna's mother carries hemophilia A is _____ .

3. The chance that Joanna is a carrier is _____ .

4. The chance that Joanna will have a son with hemophilia is _____ .

5. Measuring the amount of factor VIII in Joanna's blood is not a very accurate expression of her genotype because _____ .

6. The accuracy of the genetic marker test in establishing Joanna's carrier status depends upon _____ .

7. Describe the procedure by which bacteria are genetically engineered to produce human clotting factor VIII.

HYPERKALEMIC PERIODIC PARALYSIS

Key Words

Artificial insemination
Gene pool
Hardy Weinberg equilibrium
Homology
Inbreeding
Mode of inheritance
Pedigree analysis

The quarter horse was originally bred to run the quarter mile in the 1600s. Today 2.9 million of the animals are officially registered. They reproduce by artificial insemination so that breeders can keep records of the relationships of individuals. Breeders of domesticated animals often rely on a few males with desirable traits to inseminate large numbers of females. As a result, 30% of the current quarter horse gene pool reflects the genetic contribution of only four stallions. One particular male has fathered much of the present quarter horse population. He was selected because of his superior musculature.

Unfortunately, the popular stud brought an undesirable trait, too—episodes of weakness and paralysis that cause horses to collapse, not a terrific trait for a racehorse. The disorder is called hyperkalemic periodic paralysis (HYPP). The overdeveloped muscles have increased electrical activity that probably causes the periods of collapse. All of the affected horses descend from the original muscular stallion.

Worksheet

1. Humans have a form of HYPP with attacks of weakness and paralysis similar to those of quarter horses. This suggests that

    a. the genes for this disorder in humans and horses are highly conserved.
    b. humans have the potential to run very fast.
    c. humans are directly descended from horses.
    d. the trait is infectious to jockeys.

2. It would be difficult to construct a pedigree to trace HYPP in quarter horses because _____.

3. The current quarter horse population is not consistent with Hardy Weinberg equilibrium because

    _____.

4. Occasionally a horse is severely affected by HYPP with noisy breathing caused by periodic paralysis of throat muscles and very early onset. An explanation for the more serious symptoms in these individuals is probably

    _____.

LATTICE CORNEAL DYSTROPHY

Key Words

Evolutionary conservation   **Mutation**
Genetic code   **Pedigree analysis**
Mode of inheritance

A 72-year-old man, Otto, has very poor vision caused by a network of protein fibers that form a lattice-like blockage in his eyes. His vision began failing when he was in his thirties. The man's 42-year-old daughter Barbie developed vision loss and has the same defect as her father, the autosomal dominant condition lattice corneal dystrophy. Although symptoms generally begin after age 30, the first strands of protein in the eye can be seen by an ophthalmologist by the late teens.

A mutation in the gene for a protein called gelsolin, which binds to the contractile protein actin, causes lattice corneal dystrophy. Gelsolin clears actin away after injury or inflammation. In this family a single base change alters gelsolin as follows:

wild type MRNA . . . AGC UUC AAC AAU GCG GAC UGC UUC AUC CUG . . .
↓
AAC
↓
ser phe asn asn gly asp cys phe ile leu
↓
asn

Barbie and her husband Pat have three teenage children whom doctors have tested for lattice corneal dystrophy. The youngest and oldest, Betty and Billie, are affected, but middle sister, Bobbie, has normal eyes. Barbie's brother Rod is examined and found to be unaffected.

Worksheet

1. Construct a pedigree for this family.

2. The chance that a child of Rod's is affected is _____.

3. The result of a mutation changing the 11th MRNA nucleotide base shown, an A, to a U, would be _____ .

4. The result of a mutation changing the 3rd MRNA nucleotide base shown to a U would be _____ .

5. The mutation in this family occurs in the part of the protein that binds actin. It is a region that is highly conserved. What might be the result of a mutation in another part of the protein that is not crucial to its binding actin?

LEBER HEREDITARY OPTIC NEUROPATHY

Key Words

DNA, RNA, protein sequences
Evolutionary conservation
Mitochondrial genetics
Mode of inheritance
Mutation
Pedigree analysis

The Rogers family has several members who have been blind since early childhood. The nature of the blindness and the pedigree were not consistent with the more common vision loss inherited disorders, so part of the extended family—a young couple, Janet and Ted and their young son Alfred—visited a genetics clinic. Janet had been blind since childhood, as had her mother Rose and brother Neil; now Alfred was beginning to lose his sight. Janet's two sisters Louise and Lisa and another brother, Earl, are fine. The trait did not affect anyone else in Rose's generation nor any older relatives. The affected family members were diagnosed with Leber hereditary optic neuropathy (LHON) resulting from a mutation in a mitochondrial gene.

The mitochondrial genome consists of two rRNA genes, twenty-two tRNA genes, and thirteen polypeptide-encoding genes. The mitochondrial polypeptides form an electron transport chain that is part of the molecular machinery by which the cell extracts chemical energy from food molecules.

LHON in the Rogers family traces to a single base change in the gene encoding the final electron acceptor in the electron transport chain, a very highly conserved enzyme called cytochrome C oxidase. A change of a U to an A in the mRNA termination (stop) codon results in a lysine codon. The result is a protein with three extra amino acids: a lysine, then a glutamine, then another lysine. The three additional amino acids alter the conformation of the enzyme sufficiently to prevent it from accepting electrons. In the Rogers patients, this enzyme activity was completely gone.

Worksheet

1. What is unusual about mitochondrial inheritance?

2. Draw a pedigree for the Rogers family.

3. Two different ways that a stop codon can be altered to a lysine codon by the substitution of an A for a U are

   a.

   b.

4. Write a possible mRNA sequence for the two codons following the mutated stop codon.

5. Write a DNA sequence for the six bases following the mutated stop codon.

6. The mutation in the Rogers family is:
   a. nonsense and a transition.
   b. nonsense and a transversion.
   c. missense and a transition.
   d. missense and a transversion.

7. What, given information about cytochrome C oxidase, suggests that it would produce a serious phenotype if its gene was mutant?

8. In whom did the original mutation causing LHON most likely occur?

LI-FRAUMENI FAMILY CANCER SYNDROME

Key Words

Cancer
Genetic code
Germline mutation
Mode of inheritance
Somatic mutation
Pedigree analysis

About 100 families worldwide have the Li-Fraumeni family cancer syndrome. Affected people inherit a germline mutation in an autosomal gene called p53. Cancer develops when the second p53 gene is mutated in somatic tissue. The p53 protein normally functions as a tumor suppressor.

In one Li-Fraumeni family, 25-year-old Clint has bone cancer. He had two sisters; Martha died of breast cancer when her son David was 8 years old. Clint's other sister, Tina, died of osteosarcoma (bone cancer) at age 19. Their father died at age 27, also of bone cancer.

Geneticists sequence Clint's p53 gene and find an insertion of one extra "C" in a stretch of four C's. The result is a p53 protein shortened by 212 amino acids. Clint's young daughter Jill and nephew David, who are healthy, are tested for the mutation. They both have it.

Worksheet

1. Complete this family's pedigree by filling in the symbols of all members who have inherited the p53 germline mutation.

2. What must happen for Jill or David to develop cancer?

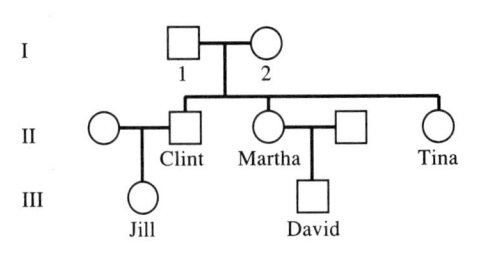

3. How can the mutation in this family, involving only one DNA base, cause as drastic a change in the gene product as 212 missing amino acids?

4. Jill and David are young children, yet their parents know that they face a very high risk of developing certain types of cancers. How can this information be used to protect their health?

5. How does the wild type version of the p53 gene function?

6. The inheritance pattern of the Li-Fraumeni cancer syndrome most closely resembles that for
   a. fragile X syndrome.
   b. retinoblastoma.
   c. Burkitt's lymphoma.
   d. cystic fibrosis.
   e. hemophilia A.

MENKES DISEASE

Key Words

Amino acid structure
Conditional probability
Mode of inheritance

Martha and Eliot D. are healthy, and in their early twenties, they have their first child Rick. The newborn is striking in appearance with short white stubby hair. Unfortunately, his hair is not the only abnormality.

At Rick's first "well baby" check-up, the physician is alarmed that the child has barely grown. The lagging growth continues. By 2 months, when babies begin to respond to their surroundings, Rick is disturbingly oblivious. His responses worsen. A series of tests reveal brain degeneration, artery abnormalities, and bending of the long leg bones.

Martha is puzzled when the physician takes pieces of hair from different parts of her head and sends them to Johns Hopkins University to be tested. When the results come in, some of Martha's hairs are unusually twisted. The doctor's suspicions are confirmed; Rick has inherited Menkes disease from his mother who is a carrier.

Menkes disease is an inborn error in copper metabolism. In several tissues, copper cannot properly exit cells. Because copper is necessary for several different enzymes to function, the deficiency of this essential trace element in many tissues causes varied symptoms. For example, copper is needed for disulfide bonds to form in keratin, the major protein in hair. The result is the characteristic twists to the hair shaft. Arteries are very weak because the connective tissue that forms part of their walls lacks crosslinks in the proteins collagen and elastin. The crosslinking reaction is catalyzed by an enzyme that requires copper. Copper is also needed for the function of an enzyme that helps the body utilize vitamin C. This explains why Menkes patients' bones resemble those of people with scurvy caused by vitamin C deficiency.

Worksheet

1. The mode of inheritance of Menkes disease is _____ .

2. Martha has a sister Emily who is thinking about starting a family. She wonders if she, too, could have a child with Menkes disease. The chance that Emily is also a carrier of Menkes disease is _____ . The chance that a son of hers would be affected, assuming that the boy's father does not have the Menkes gene, is _____ .

3. Martha's brother George is also concerned that he could father a son with Menkes disease. He and his wife Phyllis have two healthy daughters. The chance that a son of theirs inherits Menkes disease is _____ .

4. The characteristic hair of Menkes disease results from the inability of hair keratin to form disulfide (sulfur-sulfur) bridges. The two amino acids most likely to be affected are _____ and _____ .

5. The phenomenon responsible for the observation that Martha has some unusual hairs and some normal ones is

   _____ .

6. An Australian physician, D.M. Danks, saw his first patient with Menkes disease at Johns Hopkins University in 1971. He noted the striking resemblance between the patient's hair and that of sheep in his homeland that graze on copper-deficient soil. The sheep hair anomaly is a _____ of Menkes disease.

MULTIPLE ENDOCRINE NEOPLASIA

Key Words

Cancer
Genetic heterogeneity
Genetic marker
Linkage
Mode of inheritance
Segregation

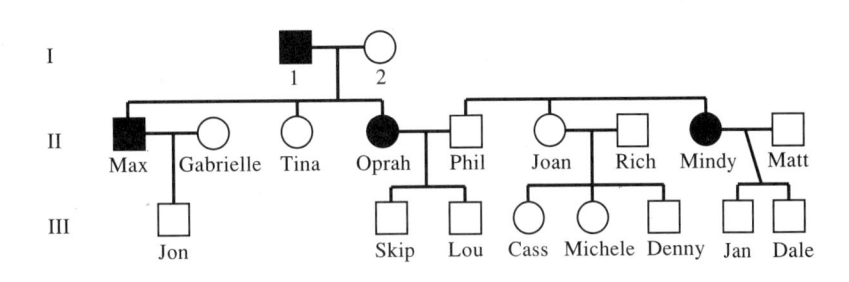

Oprah Q. sees her family physician because of severe headaches, blurry vision, bouts of sudden sweating, and rapid heartbeat. At the exam, the nurse notes that Oprah's blood pressure is high. It has been high at other visits, but sometimes it is normal. After considering the symptoms, the doctor orders blood and urine tests and is rather evasive with his patient. A few days later, Oprah is sent for an X ray of her adrenal glands to confirm what her doctor suspected—she has a small tumor in one adrenal gland. The condition is called pheochromocytoma. The tumor was found early enough to be removed, and the symptoms vanished.

Pheochromocytoma is often part of an inherited cancer syndrome called multiple endocrine neoplasia type 2 (MEN2). It affects glands of the endocrine system, particularly the adrenal glands and the thyroid. Another common manifestation is a precancerous condition of the parathyroid glands called parathyroid hyperplasia. The illustration shows the locations of these glands. The tumors and growths of MEN2 have wide-ranging bodily effects because the hormones that they overproduce affect many organs.

Oprah did not realize that she has inherited MEN2, an autosomal dominant condition with a penetrance of 80%. Her brother Max had a thyroid tumor, but she does not speak to him often and did not know of his illness. Her sister Mindy had a "mild parathyroid problem" as a teenager, but no one had mentioned MEN2. Oprah, Max, and Mindy's father had died of cancer, but by the time it had been diagnosed, it had spread to his liver. The cause of death was officially liver cancer, but it could have begun in the adrenal or thyroid glands, both of which were also affected. The doctor, knowing that pheochromocytoma often affects several family members, takes a family history and concludes that the tumors and overgrowths are related, and part of MEN2.

Worksheet

1. Why is it important for Oprah to know that her cancer is not an isolated case, but part of an inherited syndrome?

2. In one study, sixteen families were found to have a small deletion of chromosome 20p segregating with members who have MEN2 but not with healthy members. In another study of a large family, RFLP mapping implicated chromosome 10 in MEN2. These results mean that MEN is

    a. genetically heterogeneic.
    b. pleiotropic.
    c. incompletely penetrant.
    d. autosomal dominant.
    e. caused by a trisomy.

3. The chance that Jan or Dale develops an endocrine tumor or overgrowth is _____.

4. Why would a genetic marker test be a more reliable diagnostic tool for MEN2 than

   a. a family history of cancer?
   b. blood and urine tests?
   c. diagnostic scans?

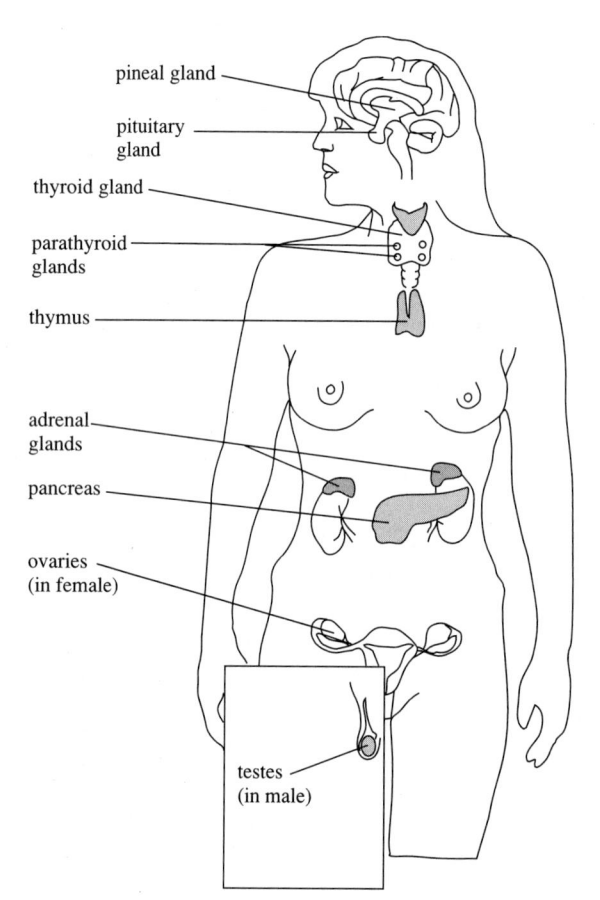

Location of the endocrine glands. The endocrine system includes several glands that contain specialized cells that secrete hormones, as well as cells scattered among some of the other organ systems that also secrete hormones.

MYOTONIC DYSTROPHY

Key Words

Control of gene expression　**Pedigree analysis**
Fragile X syndrome　**Pleiotropy**
Genomic imprinting　**Variable expressivity**
Mutation

The Rushmore family illustrates genetic anticipation for myotonic dystrophy, which means that the symptoms of this autosomal dominant condition worsen with each generation. The grandmother Elizabeth has a very mild case with muscle weakness, cataracts, and insomnia as the only symptoms. Her only child Catherine is more disabled by the condition. At age 46 she needs a cane to walk and cannot lift anything because of her weakened muscles.

Catherine's symptoms began shortly after she married Henry when she was 25 years old. Then her weakness was so minor that she did not connect it to her mother's muscle weakness, so the problem did not influence the couple's decision to have children. They had a son Scott a year later and a daughter Bonnie two years after that, and both seemed healthy as small children. Then, when Scott was 7 and Bonnie 5, Catherine and Henry had another child Jared who was born with what doctors diagnosed as congenital myotonic dystrophy. Not only were his muscles practically useless, but he also had serious heart and digestive problems. He died shortly after birth.

When the Rushmores realized that Jared, Catherine, and Elizabeth had the same illness and that it was inherited, they began to fear for the health of Scott and Bonnie. When they were teens, Bonnie developed muscle wasting—earlier and more severe than that of her mother.

Catherine asked her physician how the disease could worsen with each generation. The doctor referred her to a report in the March 6, 1992 issue of *Science*. It explained that myotonic dystrophy is associated with an increase in the number of repeats of a trinucleotide, GCT. A normal person has 5 to 27 repeats in the gene responsible for the disorder, whereas affected individuals have 50 to 1000 repeats. This many repeats somehow makes the gene unstable.

Also intriguing is the finding that all severely affected newborns, like Jared, receive the mutant gene from their mothers, and all congenitally affected individuals have affected grandmothers.

It is not yet known how the GCT repeats cause symptoms of myotonic dystrophy, but the repeated region lies within the exon closest to the 3' end of the gene. The region is transcribed into mRNA, but it is not translated into protein. The protein product of the gene is called myotonin-protein kinase, and it may function in signal transduction (transmission of biochemical information from outside to inside the cell).

Worksheet

1. Draw a pedigree for this family.

2. A disease that is more severe when inherited from a particular gender parent displays _____ _____ .

3. One study examined the family histories of twenty-three children with myotonic dystrophy who have normal parents. Two explanations for how affected children can have unaffected parents are

    a.

    b.

4. Myotonic dystrophy is pleiotropic and variably expressive. What is the difference between these two characteristics of a phenotype? How does myotonic dystrophy illustrate pleiotropy and variable expressivity?

5. How can the GCT repeats affect gene function if they are located in a part of the gene that is not translated into protein?

NEPHROLITHIASIS

Key Words

Conditional probability **Mode of inheritance**
Consanguinity **Pedigree analysis**
Degree of relationship

Over a period of two years, doctors at a medical center examined four young men with kidney stones and protein in their urine. Ned, Ted, Fred, and Jed each saw a different doctor. At a clinical staff meeting, the doctors realized that these patients had very similar symptoms. Each had been diagnosed differently probably because kidney stones are part of many syndromes. On further analysis they were all diagnosed with nephrolithiasis, an inherited disorder.

The four men all had different last names, so it had not occurred to anyone that they might be related. Ned, Ted, and Fred's mothers, however, are sisters. These sisters, Kelly, Nelly, and Ellie, also have identical twin brothers Sam and Cam who, like their sisters, are healthy. Ned, Ted, and Fred's maternal grandfather Red also had progressive kidney failure. Jed, the other young man with kidney stones, has a maternal grandmother who is a sister of Red. Jed's Uncle Ed also has the kidney ailment.

The family pedigree is:

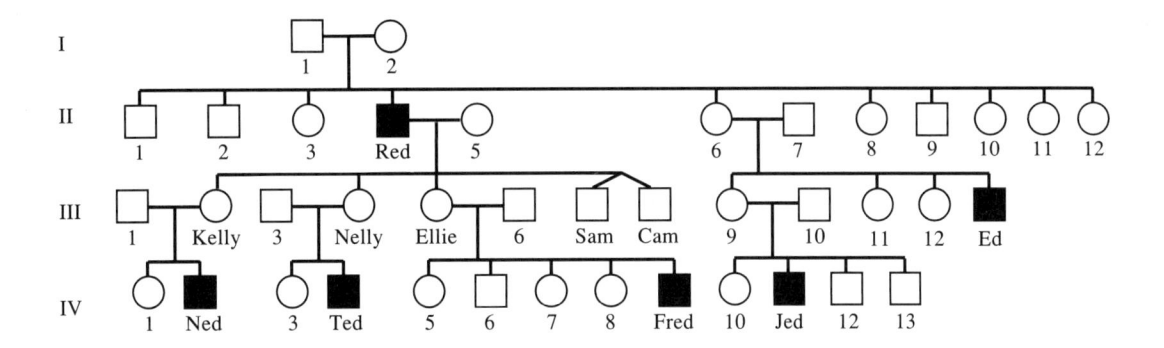

Worksheet

1. Indicate on the pedigree who must be carriers.

2. The most likely mode of inheritance is _____ because _____.

3. Treatment for nephrolithiasis is a kidney transplant. Which individuals would be the most successful donors for Ned, Ted, Fred, and Jed?

4. Jed plans to marry Elly May, Ned's sister. The risk that a daughter of theirs would inherit nephrolithiasis

   is _____, and the risk that a son of theirs would inherit nephrolithiasis is _____.

NEUROFIBROMATOSIS TYPE 2

Key Words

Conditional probability
Genetic markers
Incomplete penetrance
Mendel's first law
Mode of inheritance
Pedigree analysis
Presymptomatic diagnosis

In an extended family, several young adult members are deaf in both ears and have several brown skin marks. The seven cousins are seen at a clinic for hearing impairment. There are more than 100 forms of inherited deafness, and geneticists are eager to pinpoint which one is affecting this family.

The finding of small tumors in the fatty Schwann cells ensheathing the auditory nerve of the inner ear combined with observation of the skin marks suggest that the family members have neurofibromatosis type 2 (NF2). Further clues are the recollections that a grandfather of the young people died of a brain tumor and was deaf in both ears, and that six of the affected cousins had a parent who was either deaf, had a brain tumor, or both. Their pedigree is shown below.

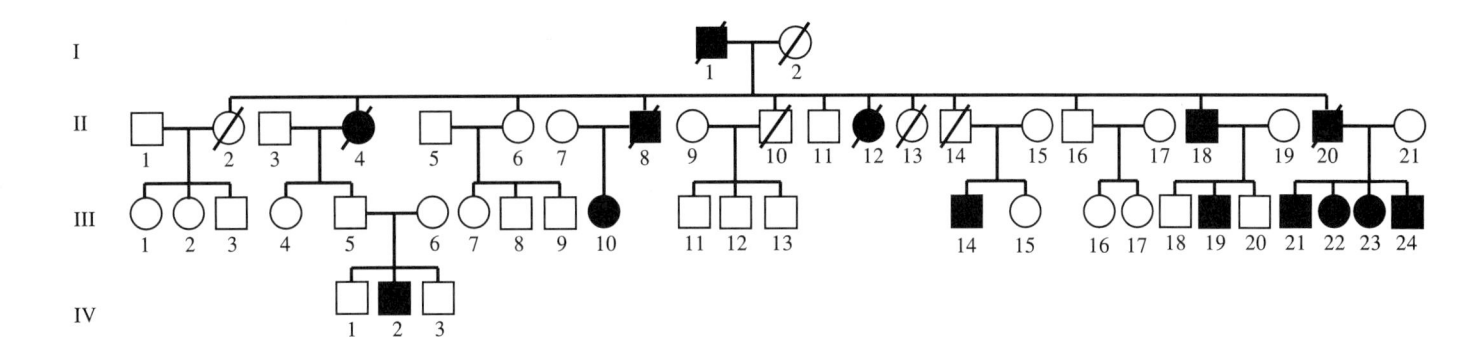

NF2 maps to chromosome 22, and it is autosomal dominant. A similar and more common disorder is NF1, which is also autosomal dominant and maps to chromosome 17. In NF1 tumors grow around peripheral nerves just beneath the skin causing bumps, and the *cafe-au-lait* skin marks are more abundant. In NF2 the tumors affect the central nervous system (the brain and spinal cord).

Because NF2 is incompletely penetrant and symptoms are sometimes not noticed until one's thirties, a genetic marker test was developed to diagnose the condition before symptoms appear. The marker is the dinucleotide CA that repeats a different number of times in people who have inherited the NF2 mutant allele, compared to individuals within a family who have not inherited the disease-causing allele. The recombination map distance between this marker and the NF2 gene locus is 10 centiMorgans.

Worksheet

1. The probability that all four children of individuals II20 and II21 inherit NF2 is _____ .

2. Two individuals in whom the NF2 mutant allele is nonpenetrant are _____ and _____ .

3. Why is presymptomatic diagnosis using the CA repeat polymorphism not 100% accurate?

4. Why would the marker based on the CA repeat not be useful to diagnose NF1?

5. Individual III10 meets a young man with NF1 at the clinic. They marry, and she becomes pregnant. The probability that their child would not inherit either form of the illness is _____.

NON-INSULIN DEPENDENT DIABETES MELLITUS

Key Words

Candidate gene	Intron
cDNA	Polymorphism
Concordance	RFLP
Environment	Twin studies
Genetic heterogeneity	

In non-insulin dependent diabetes mellitus (NIDDM), cells do not respond to the pancreatic hormone insulin that normally signals them to admit the sugar glucose from the blood. Symptoms include glucose (sugar) in the urine and increased levels in the blood, frequent urination and thirst, and often obesity. Secondary problems arise in various tissues as a result of the defect in glucose metabolism. Symptoms of NIDDM usually appear after age 40, but biochemical signs may be present earlier.

Despite the high prevalence and ability to control symptoms of NIDDM, we still do not know the underlying cause of the disorder. We know that heredity plays a role because concordance in monozygous twins is 90%; but according to an editorial in The New England Journal of Medicine, "studies of identical twins with NIDDM have clearly shown that the disease is genetic, although the pattern of inheritance has defied classification." Another piece of genetic evidence is that a person with one parent who has NIDDM has a 40% risk of developing the condition. One reason that NIDDM cannot be easily categorized into a Mendelian mode of inheritance is that the environment plays a role in terms of diet.

The search for a candidate gene behind NIDDM focuses on known proteins whose functions could explain the inability to respond to insulin. Clues come from the characteristics of NIDDM at the biochemical and cellular levels. In this disorder, muscle and liver cells are unable to convert glucose into the complex carbohydrate glycogen. A key enzyme in this conversion is glycogen synthase, which insulin activates. Once inside a cell, glucose is either immediately metabolized to carbon dioxide and water or stored as glycogen, as follows:

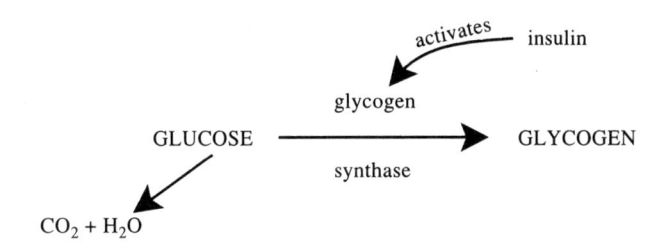

$$\text{GLUCOSE} \xrightarrow{\text{glycogen synthase}} \text{GLYCOGEN}$$

with insulin activating glycogen synthase, and an alternative pathway $\text{GLUCOSE} \rightarrow CO_2 + H_2O$.

In people who have NIDDM and in some of their unaffected first degree relatives, glycogen synthase does not respond to insulin. This symptom is termed "insulin resistance." The gene for glycogen synthase on chromosome 19 is, therefore, a candidate gene for NIDDM.

Danish researchers synthesized a cDNA probe corresponding to most of the glycogen synthase gene. By cutting the genes from patients and a control group of nondiabetics with the restriction enzyme XbaI, followed by Southern blot analysis, the researchers identified two polymorphisms in an intron of the gene designated A1 and A2. Results were as follows:

- 30% of 107 patients with NIDDM have the A2 allele.
- 8% of 164 nondiabetic patients have the A2 allele.

Furthermore, those diabetics with the A2 allele have a stronger family history (more affected relatives), double the risk of hypertension (high blood pressure), and have more severe insulin resistance than diabetics lacking the A2 allele. The researchers conclude that the A2 allele can be used to identify a more severely affected subgroup of people with NIDDM.

Worksheet

1. Why does the Danish research not prove that the glycogen synthase gene is a direct cause of NIDDM? What other kinds of evidence would support this role for the gene?

2. The cutting site for XbaI in one allele is CCTAGA and TCTAGA in the other. The reason that the amino acid encoded by this sequence is not relevant is that _____ .

3. If an identical twin develops NIDDM, the chance that the other twin will not is _____ .

4. Give two explanations for the observation that not all patients with NIDDM have the A2 allele for the glycogen synthase gene.

5. How did the researchers obtain a cNDA for glycogen synthase?

6. NIDDM is probably
    a. genetically heterogeneic.
    b. variably expressive.
    c. incompletely penetrant.
    d. multifactorial.
    e. all of the above.

OSTEOPETROSIS

Key Words

Consanguinity **Mode of inheritance**
Incomplete penetrance **Pedigree analysis**

Bone is a living tissue that is constantly being remodeled with tissue being added and degraded so that the bone maintains its shape while enlarging as a person grows. Osteopetrosis is a disorder in which bone is not resorbed properly in the remodeling process. The marrow within the bones, which contains blood cell precursors, is also abnormal.

There are two forms of osteopetrosis. The Neumans have a variant of the disorder that is inherited as an autosomal dominant trait that is incompletely penetrant. Victor Neuman's back trouble was recently diagnosed as a compressed vertebra, one sign of osteopetrosis. His father John had very fragile bones and frequent dental abscesses, the two other symptoms of the disorder. Victor's sister Ashley often breaks bones when she takes falls skiing that would not harm other people, so she may be affected too.

Victor and Ashley's brother Jack is very healthy, but Jack and Nikki's son and daughter, Nicholas and Victoria, have each had several dental abscesses, and each has broken an arm. Ashley is married to Ryan, and their daughter Drucilla and son Neil, so far, appear to be healthy.

The Barbers have the second, more serious form of osteopetrosis that is inherited as an autosomal recessive trait. This form is very rare, and when seen, it is usually in an isolated population or in a person whose parents are related by blood. In severe osteopetrosis, the disruption in bone remodeling is pervasive during fetal existence, and before birth there is already life-threatening anemia caused by the abnormal bone marrow and a greatly enlarged spleen and liver. The child becomes deaf and blind shortly after birth. Death comes in early childhood.

Charlie and Midge Barber have one healthy son, but have lost two daughters to the condition. Their pedigree is shown below:

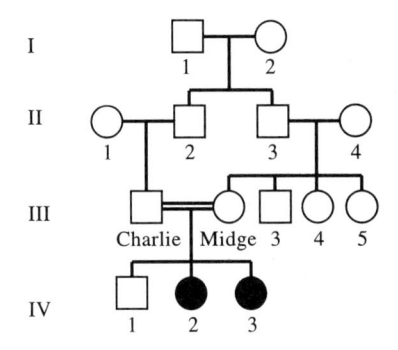

Worksheet

1. How is osteopetrosis variably expressive?

2. The fact that osteopetrosis exists in autosomal dominant and autosomal recessive forms means that it is
   _____.

3. Draw a pedigree for the Neumans indicating known carriers.

4. We know that _____ Neuman is non-penetrant because _____.

5. What fact about the Barber family has contributed to their children's inheriting this rare condition?

6. How can a pedigree for a condition that is autosomal dominant and incompletely penetrant resemble one for a condition that is autosomal recessive?

PELIZAEUS-MERZBACHER DISEASE

Key Words

Genetic code
Genotype
Mutation
Pedigree analysis

Phenotype
Polymorphism
RFLP mapping
Wild type

Philip and Wendy are two healthy 25 year olds when they have Danny. He seemed normal in early infancy, smiling and feeding well. At 10 weeks he developed stridor, a shrill sound upon inhaling that suggests a blockage in the larynx. Wendy took him to the emergency room where the doctor said his epiglottis (the trapdoor-like structure separating the esophagus from the trachea) was floppy, a common condition usually outgrown by 2 years. Danny's stridor soon improved on its own, but his eyes began to wander, a symptom called nystagmus.

As the months passed, Philip and Wendy became more concerned about Danny's lack of progress. When at 13 months the boy still could not sit up, his pediatrician ordered a computed tomography (CT) brain scan. This showed minor degeneration of nerve cells in the cerebrum, the upper part of the brain controlling reasoning, sensation, and perception.

Because the brain involvement was minor, and Danny's karyotype and metabolic tests were normal, a diagnosis was not made. Unfortunately, no one took a complete medical history that would have revealed the important information that Wendy's brother Peter had died at a year of age. All her parents could recall of Peter's illness was that his eyes wandered and he simply stopped developing.

When Danny was 21 months old, Wendy gave birth to Christopher. Three months later, Danny had a magnetic resonance imaging (MRI) brain scan that showed widespread demyelination—loss of the fatty sheath that surrounds nerve cells and enables them to rapidly transmit electrochemical messages. At about this time, Chris developed the snuffly breathing sound that had first alerted the parents to Danny's condition. Now that a second child had the same symptoms, a geneticist was consulted. A month later, Danny died of respiratory failure.

Both boys were diagnosed with Pelizaeus-Merzbacher disease caused by a gene on chromosome Xq. DNA from the boys, their mother, maternal aunt Vicki, and maternal grandmother was sequenced in the region of the gene that encodes a membrane protein called proteolipid protein. A point mutation altering an A to a C at nucleotide position 541 was detected in the boys, their mother, and grandmother, but not in their aunt. This mutation changes the amino acid threonine (thr) to proline (pro).

Worksheet

1. Construct a pedigree for this family.

2. The mode of inheritance for Pelizaeus-Merzbacher disease in this family is _____ _____.

   However, in some families with identical symptoms, the mode of inheritance is different, and the proteolipid

   protein gene on the X chromosome is wild type. This means that this disorder is _____ _____.

3. For Pelizaeus-Merzbacher disease the gene product cannot be measured to indicate carrier status. A way to genetically detect carriers is to screen for a polymorphism that is found in the mutant proteolipid protein gene. This RFLP is a silent transition mutation in codon 202, which is GAU in the wild type. What is the codon in the mutant gene?

4. Why is the mutation in codon 202 not responsible for the phenotype?

5. The codon change at nucleotide position 541 and amino acid 181 that causes the phenotype is a _____ to a _____ . Is this a transition or a transversion?

6. Does Aunt Vicki need to worry about passing on this illness to her children? How do you know this?

7. What are the chances that a daughter of Wendy and Philip would inherit Pelizaeus-Merzbacher disease?

PLACENTAL AROMATASE DEFICIENCY

Key Words

Cancer
Environment
Exon
Frameshift mutation

Gene expression
Intron
Mutation
Phenotype

Aromatase is an enzyme that catalyzes conversion of androgens (male sex hormones) to estrogens (female sex hormones). The aromatase gene is highly expressed in the ovaries and in the placenta, the organ through which a pregnant woman and fetus exchange nutrients and wastes. The gene is expressed to a lesser extent in liver, muscle, adipose cells, and hair follicles where the enzyme functions as a growth or differentiation factor. Aromatase is also expressed in the brain where it may play a role in establishing sexual identity.

People with aromatase deficiency are extremely rare. One case recently arose in a pregnant woman. She had been fine before pregnancy, but as pregnancy progressed, she became more and more masculine in appearance despite her bulging middle. The newborn looked male, but the karyotype was XX. The placenta was found to be completely devoid of aromatase.

The aromatase gene of the mother was examined. An 87 base insert was found in the mRNA, and the protein had 29 extra amino acids, a drastic enough change to render the enzyme inactive. The gene appeared to be activated only during pregnancy—the mRNA was no longer found after the birth. The infant had two wild type alleles for the autosomal recessive aromatase gene.

Worksheet

1. Could an 87 base insertion disrupt the reading frame? Why or why not?

2. An insert of genetic material could be caused by

   a. a nonsense mutation in the gene.
   b. a missense mutation within an intron.
   c. a point mutation at a splice junction causing an exon to be translated.
   d. a point mutation at a splice junction causing an intron to be translated.

3. How can the infant have a phenotype suggesting estrogen deficiency if she did not inherit the mutant alleles?

4. Some breast cancer cells bear receptors for estrogen, and are stimulated to divide when estrogen is present. The drug tamoxifen works by blocking estrogen receptors on breast cancer cells. Knowing that aromatase is necessary for synthesizing estrogens, suggest treatment approaches at the nucleic acid and protein levels to intervene with the expression of aromatase to prevent or control the spread or recurrence of estrogen sensitive breast cancer.

5. Hypothesize two explanations for why people with aromatase deficiency have hardly ever been seen.

6. How do we know that the aromatase deficiency in the pregnant woman was activated only during pregnancy?

PRADER-WILLI/ANGELMAN SYNDROMES

Key Words

Meiosis
Nondisjunction
Segregation

Prader-Willi and Angelman syndromes are forms of mental retardation that are parts of distinct syndromes. The Prader-Willi child eats uncontrollably, becoming obese, and is short with small hands and feet and almond-shaped eyes. Gonads are underdeveloped, and the child is poorly coordinated. Angelman syndrome is also called "happy puppet syndrome" after the appearance of the child. He or she laughs inappropriately and frequently, walks oddly, has puppetlike limb movements, a large jaw, cannot speak, and has a small head with an open mouth and protruding tongue.

An absence of genes or gene function in the same region of chromosome 15 can cause either syndrome; 70% of Prader-Willi patients and 80% of Angelman patients have tiny deletions in both chromosome 15s. Molecular studies on the non-deletion cases, however, show surprising results. Prader-Willi syndrome is associated with inheriting two copies of this chromosomal region from the mother, an example of uniparental disomy. Angelman syndrome is associated with two paternal copies of the same chromosome region. Geneticists are still rather confused about these two syndromes.

Worksheet

1. Whether a child has Prader-Willi or Angelman syndromes not caused by chromosomal deletion seems to depend upon whether the responsible gene is inherited from the mother or the father. This is an example of

   _____   _____ .

2. How might nondisjunction account for the uniparental disomy seen in Prader-Willi and Angelman syndromes?

3. How does uniparental disomy in Prader-Willi and Angelman syndromes contradict Mendel's first law?

PROTEIN C DEFICIENCY

Key Words

Environment
Immunogenetics
Mode of inheritance
Pedigree analysis
Wild type

In protein C deficiency, patients lack part of a blood plasma glycoprotein that regulates blood clotting. Small blood vessels leak throughout the body producing bruises and widespread skin lesions called purpura fulminans.

An infant, Ian, is admitted to the hospital with red marks on his fingers and legs that turn large and blue. He is diagnosed with protein C deficiency because several years earlier, his sister Keri died from it. He is given protein C pooled from blood donations. He recovers.

To trace the disorder in his family, Ian's parents and siblings are tested for the level of an antigen indicating the amount of protein C. Since this is an autosomal recessive condition, it is expected that the parents are carriers, and, therefore, would have half the normal amount of the antigen and protein C. An antigen value of 1.00 indicates normal (wild type). The pedigree and antigen values are:

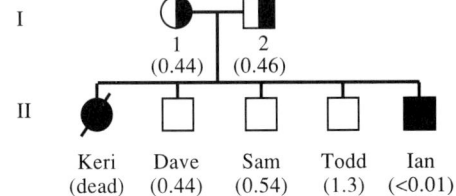

I: 1 (0.44) — 2 (0.46)
II: Keri (dead), Dave (0.44), Sam (0.54), Todd (1.3), Ian (<0.01)

Worksheet

1. For each of Ian's brothers, indicate whether he is a carrier of protein C deficiency, has the syndrome, or has two wild type alleles.

    a. Dave
    b. Sam
    c. Todd

2. If Ian grows up and has children with a woman who is wild type for the protein C gene, the risk that each child of theirs would be a carrier of this disorder is _____.

3. Purpura fulminans can also be caused by leukemia, rubella (German measles), or taking certain drugs. Purpura fulminans caused by any of these sources is a _____ of inherited protein C deficiency.

4. Obtaining protein C from donated plasma could introduce infection. A different source of protein C would be from _____.

PSEUDOHERMAPHRODITISM

Key Words

Consanguinity
DNA to RNA to protein
Genetic code
Genetic drift
Genotype

Mutation
Pedigree analysis
Penetrance
Phenotype

In twenty-nine families in the Dominican Republic, a single base change in a gene encoding the enzyme 5-alpha-reductase renders genital tissues in males unable to respond to the male sex hormone testosterone. The mutant allele is recessive and located on chromosome 2. Homozygous recessive females have a normal phenotype.

Homozygous recessive males have a normal male chromosome constitution of XY, but they have ambiguous genitalia, and may actually look more like a girl than a boy. Their testicles are undescended, and they often have a vagina-like pouch near a very small penis. They are locally called "guevodoces," which means "penis at twelve" in Spanish. At puberty, they develop some male secondary sexual characteristics, such as a deep voice, heavy musculature, and a larger penis. Before the condition was understood, affected individuals were raised as girls until puberty, and then as boys.

All twenty-nine affected families have the same mutation, a change in the RNA codons for the 246th amino acid from CGG to UGG. There is much consanguinity, and their society is geographically and socially isolated.

Worksheet

1. Why would it be difficult to construct a pedigree for this condition?

2. The amino acid change that causes pseudohermaphroditism is _____ to _____ .

3. The DNA sequence of the region affected by the mutation is _____ .

4. Three reasons why this disorder is prevalent in this population are:

    a.

    b.

    c.

5. The fact that all twenty-nine families have the same mutation suggests that _____ .

6. The fact that the phenotype is not expressed in females illustrates

    a. sex influenced inheritance.
    b. sex linked inheritance.
    c. sex limited inheritance.
    d. genomic imprinting.
    e. sexual harassment.

7. Why would testosterone shots not help young, affected boys develop normally?

8. Is this form of pseudohermaphroditism incompletely penetrant? How do you know this?

PURINE NUCLEOSIDE PHOSPHORYLASE DEFICIENCY

Key Words

Antibodies	Enzyme function
B cells	Immunogenetics
DNA structure	Phenocopy
Environment	T cells

The immune system is built of several varieties of white blood cells called T cells and B cells. T cells secrete biochemicals that attack cancer cells and pathogens. T cells called helpers activate B cells, which manufacture antibody proteins that engulf and destroy nonself cells including pathogens.

T cell function is very complex, and, therefore, there are many ways that immunity can be impaired. Immune dysfunction affecting T cells and B cells is called severe combined immune deficiency (SCID). One rare form of SCID was initially described in 1975 in a 5-year-old girl who lacked T cells and had too few B cells. Her B cells were being depleted because her immune system was attacking them (autoimmunity). The cause of the girl's symptoms of recurrent infections and poor weight gain was traced to lack of an enzyme, purine nucleoside phosphorylase (PNP).

PNP is a trimer of identical polypeptide subunits, each with a substrate binding site. It frees guanine from its sugar and phosphates. When PNP is absent in T cells, deoxyguanosine triphosphate (dGTP) accumulates. The accumulation inhibits an enzyme called ribonucleotide reductase. Inactivity of this enzyme blocks DNA synthesis. Because the immune response depends upon rapid division of T and then B cells, lack of PNP ultimately suppresses immune function.

PNP deficiency is recessive and is caused by a mutant gene on chromosome 14. It is 9 kilobases long and includes six exons. The mutant allele differs from the wild type allele by a single base change that results in an amino acid substitution.

Since 1975 thirty-two other children with PNP deficiency have been described. In addition to infections and failure to thrive, some of the children have autoimmune disorders, neurological impairment, or cancer. They generally die before age 20 of infection or cancer.

Worksheet

1. If lack of PNP blocks DNA synthesis, which is a prerequisite for cell division (cancer is the consequence of too frequent cell division), why are children with PNP deficiency at high risk for developing cancer?

2. The phenotype corresponding to a deficiency of ribonucleotide reductase would be _____.

3. The mutation that causes PNP deficiency is

    a. nonsense.
    b. missense.
    c. frameshift.
    d. insertional.
    e. antisense.

4. A PNP subunit is much smaller than corresponds to the size of the gene that specifies its synthesis. This is

    because of _____.

5. Why is interference with helper T cells particularly threatening to health?

6. A phenocopy of severe combined immune deficiency is the infectious disease _____ .

7. How is variable expressivity of PNP deficiency dependent on the environment?

RED-GREEN COLORBLINDNESS

Key Words

Mode of inheritance **Pedigree analysis**
Monozygous twins **X-inactivation**

In the pedigree below, Daren, Sharon, Jim, and Ben are red-green colorblind, a recessive trait transmitted on the X chromosome.

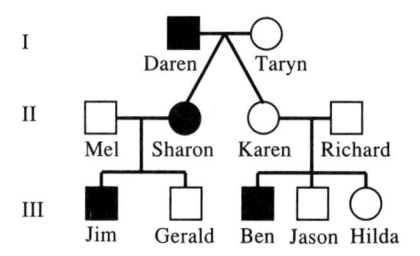

I Daren — Taryn

II Mel — Sharon Karen — Richard

III Jim Gerald Ben Jason Hilda

Worksheet

1. How can Sharon be colorblind, but Karen, her identical twin, not be? The condition is inherited.

2. Taryn is wild type for the colorblindness alleles. What are the genotypes of Sharon and Karen for red-green colorblindness?

3. What is the evidence that Karen is a carrier for the colorblindness gene?

SCHNECKENBECKEN DYSPLASIA

Key Words

Consanguinity
Mode of inheritance
Pedigree analysis

Stuart and Gloria have had four healthy children, five stillborn dwarves, and three miscarriages. In their most recent pregnancy, ultrasound showed dwarfism in the third month. The phenotype includes very short leg bones, small vertebrae, and a twisted pelvis. *Schneckenbecken* is German for "snail pelvis." The pedigree is shown below.

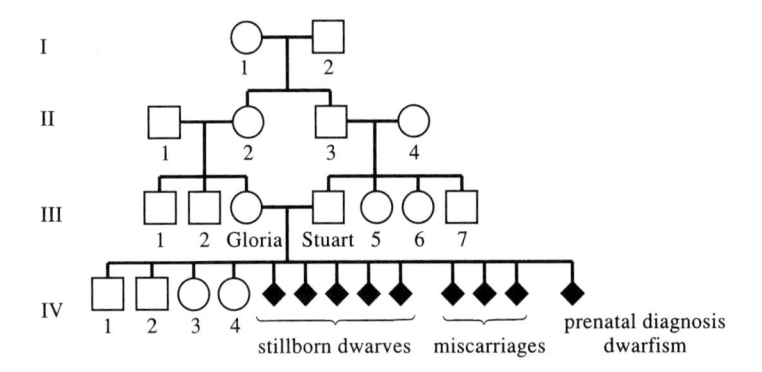

Worksheet

1. Stuart and Gloria have 1/____ of their genes in common.

2. The chance that individual IV2 is not a carrier is _____.

3. Which individuals could be carriers of this condition?

SEVERE CHILDHOOD AUTOSOMAL RECESSIVE MUSCULAR DYSTROPHY

Key Words

Genotype
Immunogenetics
Mode of inheritance
Monoclonal antibodies
Pedigree analysis
Phenotype
X inactivation

In North Africa an autosomal recessive form of muscular dystrophy is common. The symptoms are identical to those of the more familiar sex-linked Duchenne muscular dystrophy (DMD)—gradual wasting away of skeletal (voluntary) muscles, leaving the child wheelchair bound by age 10, often proving lethal by age 20.

DMD is caused by abnormal or absent dystrophin protein. This protein, along with several associated proteins, binds muscle cells to each other, enabling them to function coordinately. Without dystrophin, the entire complex of muscle proteins falls apart. In the severe childhood autosomal recessive muscular dystrophy (SCARMD) of North Africa, dystrophin is normal, but one of the associated proteins is absent.

The Kardassian family moved to Los Angeles from Egypt 8 years ago. They have two boys, Nick and Peter, who have severe muscular dystrophy, and a healthy daughter Lisa. The boys have been diagnosed with DMD based on their phenotypes, but confirmatory genetic tests find that they make normal dystrophin. Neither does their mother produce dystrophin in some cells but not in others, as carrier mothers of DMD do. The doctors are puzzled.

Nick and Peter's mother Bertha learns that her brother Lyle's daughter Serina has just been diagnosed with muscular dystrophy too. Bertha and Lyle are the only children of Arthur and Celeste who are healthy.

Worksheet

1. Intrigued by this unusual mode of inheritance for muscular dystrophy, Nick and Peter's physician sends samples of their muscle tissue to researchers who apply a radioactively labelled monoclonal antibody made in mice against human dystrophin. They compared the results on a healthy control, a boy with DMD, and Nick and Peter:

   healthy control — Muscle cells are outlined by the labelled dystrophin, which holds the muscle proteins and cells together.

   DMD boy — No cell outlines.

   Nick & Peter — Only some staining, but dystrophin is present.

   Can you explain this result?

2. Evidence that Nick and Peter's muscular dystrophy is inherited as an autosomal recessive and not a sex linked recessive trait is that _____.

3. The fact that female carriers of DMD have some cells that manufacture dystrophin and some that do not illustrates the phenomenon of _____.

4. Draw a pedigree for the Kardassians and their cousin Serina's family.

5. The risk that Lisa is a carrier for SCARMD is _____.

SICKLE CELL DISEASE

Key Words

Development	Multifactorial inheritance
Evolution	Mutation
Gene therapy	Oncogene
Globin genes	Polygenic inheritance

Hemoglobin is the iron-protein complex that carries oxygen in the blood. The protein portion consists of four subunits, two alpha and two beta. The two types of subunits are encoded by the alpha and beta globin genes, respectively. Because of differing oxygen concentration requirements during prenatal development, the fetus has two gamma globin polypeptide chains instead of the two adult beta chains. A few months after birth, the genes encoding gamma globin turn off, and the genes encoding beta globin turn on.

Sickle cell disease results from a point mutation in the gene encoding beta globin. When oxygen is low, the mutant hemoglobin polymerizes, bending the red blood cell containing it into a sickle shape. The odd-shaped red blood cells lodge in narrow capillaries causing blood to build up and intense pain.

One promising approach to treating sickle cell disease is to reactivate the fetal globin genes. This idea came from several observations:

- Some infants whose mothers were diabetic during pregnancy have abnormally high plasma concentrations of a biochemical called butyrate that prevents gamma globin from switching off and beta globin from switching on. Despite this biochemical quirk, these children are healthy.
- People with a condition called "hereditary persistence of fetal hemoglobin" are phenotypically normal.
- Saudi Arabian and Indian people who have sickle cell disease have a variant that is much milder than that seen in other populations, sometimes without any symptoms. Twenty-five percent of their hemoglobin is of the fetal variety.

The first treatment to activate fetal globin genes to treat sickle cell disease was a drug called hydroxyurea, an approach first tried in 1983. However, this drug is toxic, and the long-term use necessary to raise fetal hemoglobin levels and maintain them sufficiently to keep too many red blood cells from sickling could cause cancer, a low white blood cell count, or damage chromosomes.

In 1992, researchers found that alternating hydroxyurea with another drug, erythropoiten (EPO), works better than either treatment alone. EPO is normally made in the kidneys and helps the body to replenish red blood cell supplies when the person has anemia. It is used as a drug, manufactured with recombinant DNA technology, to treat dialysis patients, who often develop severe anemia. The hydroxyurea/EPO regimen is supplemented with dietary iron to provide the iron needed to build more hemoglobin. The three treatments boost the percentage of red blood cells with fetal hemoglobin significantly, while relieving the terrible joint and bone pain of sickle cell disease.

Another way to turn on fetal hemoglobin is to give butyrate, a natural fatty acid that, like EPO, is used to stimulate red blood cells to differentiate from their precursor forms, called reticulocytes. Butyrate given to sickle cell disease patients significantly increases the amount of mRNA for gamma globin in reticulocytes. Reticulocytes are affected because mature red blood cells lack nuclei.

Worksheet

1. Which of the following hereditary conditions might also be treated by reactivating fetal hemoglobin?

    a. Hemochromatosis is when too much dietary iron is absorbed in the intestinal tract. The condition is treated by removing blood.
    b. Hemophilia A is when a mutation in the X linked gene encoding clotting factor VIII impairs blood clotting.
    c. Beta thalassemia is when a deletion in the beta globin gene causes anemia.
    d. Hemoglobin C, with symptoms similar to those of sickle cell disease, is caused by a different point mutation in the beta globin gene.
    e. Chronic myelogenous leukemia is a cancer of certain white blood cells associated with a translocation that activates a protooncogene, leading to loss of cell cycle control in these cells.

2. The genes encoding gamma globin reside near the beta globin genes in a cluster on chromosome 6. How might these genes have arisen?

3. How can fetal hemoglobin be manufactured in a person if he or she is no longer a fetus?

4. What evidence suggested that butyrate might promote synthesis of fetal hemoglobin?

5. What would be a problem involved in simply injecting sickle cell patients with fetal hemoglobin?

6. Why can red blood cells not be genetically engineered?

7. Would you consider reactivation of fetal hemoglobin a form of gene therapy?

8. Anemia is a feeling of great fatigue resulting from too few red blood cells or from red blood cells carrying too little hemoglobin. Why is anemia both polygenic and multifactorial?

TANGIER DISEASE

Key Words

Familial hypercholesterolemia
Gene therapy
Genotype

Mendelian inheritance
Phenotype
Population genetics

Tangier disease is prevalent among residents of Tangier Island in the Chesapeake Bay. Most of these people are related to the original settlers who arrived in 1686. The disease is also called analphalipoproteinemia after its biochemical characteristic—low blood levels of alpha-lipoprotein, which normally transports cholesterol in the blood. It is inherited as an autosomal recessive trait. Sufferers of Tangier disease have low blood cholesterol but excess cholesterol in other areas, such as the thymus gland and scavenger cells of the immune system called macrophages. Other, unexplained symptoms include:

- large, orange tonsils
- inability to sense pain, heat, and cold on the skin
- wasting of muscles in the hands
- enlarged liver, spleen, and lymph nodes

Ronald is a resident of Tangier Island who has Tangier disease but feels relatively healthy. He marries Nancy, a neighbor whose mother has Tangier disease but whose father does not. Nancy does not have Tangier disease, but she has inherited the heterozygous form of familial hypercholesterolemia (FH) from her father, who died at a young age of a heart attack. In FH, an autosomal dominant condition, liver cells have half the normal number of low density lipoprotein receptors, and as a result, cholesterol accumulates in the blood causing early heart disease. Nancy knows that she has probably inherited heterozygous FH because her blood cholesterol is dangerously high without medication and strict dietary control.

Worksheet

1. Assuming that Tangier disease and FH are transmitted on different chromosomes, the chance that a child of Nancy and Ronald inheriting both disorders is _____.

2. The phenotype resulting from inheriting both disorders is probably _____ because _____.

3. Even though we know the gene and its protein product for Tangier disease, gene therapy would be difficult because _____.

4. The portion of the alpha lipoprotein molecule that is directly affected by the mutant gene causing Tangier disease is _____ .

5. A genetic phenomenon that explains why Tangier disease is prevalent on this island but very rare elsewhere is _____ .

TAY-SACHS DISEASE

Key Words

Exon Intron
Genetic code Mutation
Genetic heterogeneity

Tay-Sachs disease is an autosomal recessive condition that causes progressive nervous system degeneration. A child is deaf and blind by one or two years and usually dies by age three. The disorder is very rare in the U.S. because of screening programs to identify carriers. Couples who are both carriers often choose to avoid the birth of an affected child because there is no treatment. Worldwide, Tay-Sachs is found among Ashkenazi Jews, French Canadians, the Pennsylvania Dutch, Cajuns, and Moroccan Jews.

More than thirty mutations are recognized in the HEXA gene on chromosome 15 that causes Tay-Sachs disease. The wild type allele encodes the alpha subunit of hexosaminidase, a dimeric protein that also has a beta subunit, encoded by a different gene.

The five people below have different HEXA mutations:

- Patient A has a 4 base pair insertion in an exon.
- Patient B has a G to C mutation in the splice site of intron 12, which creates a recognition site for the restriction enzyme DdeI.
- Patient C has a deletion that eliminates a phenylalanine amino acid from the protein product at amino acid position 304. Position 305 is also a phenylalanine.
- Patient D has a C to G mutation at amino acid 180, altering a UAC to a UAG in the mRNA.
- Patient E has a G to A mutation at amino acid position 170, altering an arginine (arg) to a glutamine (gln).

Worksheet

1. Patient D's hexosaminidase A is abnormal in that it _____.

2. Two codon changes that could account for patient E's substitution of a gln for an arg, involving a G to A transition mutation, are _____ to _____ and _____ to _____.

3. The number of DNA bases deleted in patient C is _____.

4. How might Patient B's mutation alter the gene product?

5. How might another genetic condition cause symptoms similar to those of Tay-Sachs disease?

6. Patient _____ has an altered reading frame for HEXA.

7. Suggest two reasons why Tay-Sachs disease is more prevalent in certain ethnic groups.

TOURETTE SYNDROME

Key Words

Expressivity
Mode of inheritance
Pedigree analysis

Penetrance
Pleiotropy

Mike and Carrie Riley had always attributed to simple mischievousness the fact that four of their six children were considered troublemakers in school. Greg and Ken had the annoying habit of often repeating what was said to them, and both boys had been punished for suddenly yelling out obscenities in school. Alicia and Kate didn't yell, but they just could not sit still in school, always fidgeting, jerking their heads, blinking, and grimacing. So far, Joe and Cindy had no such mannerisms. Since their father Mike and his mother Rita also made these movements, everyone thought that the children had just picked up their father and grandmother's behavioral quirks from watching them.

Mike's niece and nephew, Edie and Steve, children of his sister Suzanne and her husband Adam, are generally healthy, but have peculiar habits. Edie takes about twenty showers a day and washes her hands dozens of times a day. Steve has dyslexia and also has great difficulty solving mathematical problems, such as balancing the checkbook and comparative shopping at the supermarket. Their third child Jenny is only two years old, and, so far, her behavior is normal.

Carrie's sister Vicki is in medical school. When she learns about Tourette syndrome, she immediately thinks of her troublesome nieces and nephews.

Tourette syndrome was first described by French neurologist Georges Gilles de la Tourette in 1825. Tourette's patient, Marquise de Dampierre, was a noble woman who had facial tics, yelled foul language (coprolalia) and repeated what was said to her (echolalia). Today it is known that Tourette syndrome is inherited as an autosomal dominant trait, with incomplete penetrance, and may affect as many as 1 in 200 people. However, only 1 in 10 affected individuals has symptoms severe enough to require medical attention.

The definition of Tourette syndrome has recently been expanded to include the following other disorders:

1. sleep disorders
2. very impulsive behavior
3. learning disabilities such as dyslexia and perceptual problems
4. hyperactivity and/or attention deficit disorder
5. obsessive-compulsive traits (performing a task or activity many times)

Worksheet

1. Draw a pedigree of the extended Riley family indicating individuals who probably have Tourette syndrome according to the recently revised definition.

2. The risk that Jenny has inherited Tourette syndrome is _____.

3. The facts that there are many symptoms of Tourette syndrome and different people are affected to a different degree mean that the phenotype is
    a. non-penetrant and variably expressive.
    b. pleiotropic and variably expressive.
    c. autosomal recessive and sex limited.
    d. multifactorial and sex linked dominant.

4. The risk that Greg's children will inherit Tourette syndrome is _____.

TURNER SYNDROME

Key Words

Cytogenetics
Mosaicism
X inactivation

Most people with Turner syndrome have one X chromosome and no Y chromosome in each somatic cell. About 15% of Turner patients have a different chromosomal anomaly, an isochromosome for the X, in which the chromosome consists of two long arms but no short arm. A smaller percentage of patients have yet another aberration occurring in only some of their cells: a small ring chromosome derived from X chromosome material.

In a study at Henry Ford Hospital in Detroit, 190 Turner patients were evaluated cytogenetically. Five people were found to have an X-derived ring in some cells, but other cells were XO. All of these five patients were among the 6.3% of Turner patients who are mentally retarded. The researchers hypothesized that the ring chromosome may cause mental retardation because it lacks the locus that normally inactivates one X chromosome in female cells. Over-expression of some X linked genes, somehow, causes the mental retardation.

Worksheet

1. The symptoms of Turner syndrome are _____
   _____.

2. What sex is a Turner patient?

3. An isochromosome would be expected to cause symptoms because _____.

4. One explanation for why no patients had the X-derived ring in every cell is that _____
   _____.

5. About 95% of Turner conceptions end as spontaneous abortions. Would you expect the frequency of Turner individuals with X-derived rings among spontaneous abortions to be higher or lower than that among Turner individuals who survive? Why or why not?

6. How can DNA probe technology be used to determine if an X-derived ring lacks the inactivation locus?

VESTIBULAR SCHWANNOMA

Key Words

Germline mutation **Penetrance**
Mode of inheritance **Somatic mutation**
Pedigree

Over a period of a few months, 24-year-old Tiffany noticed problems with her hearing. Not only was there a constant background noise (a condition called tinnitus), but her hearing was distorted, too. When she began to lose her balance, she had a medical exam. Her physician sent her for a magnetic resonance imaging (MRI) scan, which found tiny benign tumors resembling bunches of grapes on each of her paired eighth cranial nerves. These nerves serve the inner ear where hearing and balance are centered. The tumors occur in Schwann cells, the fatty cells that wrap around nerve cells and are necessary for fast nerve conduction.

Although Tiffany's tumors were benign, they had to be removed because if allowed to grow, they would compress vital structures. This would eventually cause facial pain and numbness, difficulty speaking and swallowing, nausea and vomiting, lethargy, coma, and death.

Tiffany has vestibular Schwannoma. She is one of the 5% of patients who have the inherited form that causes tumors in both ears resembling bunches of grapes. The 95% of patients who have the non-inherited, or sporadic form of the illness have tumors in only one ear that form a single mass. Sporadic cases are caused by two somatic mutations occurring only in the affected cells.

Tiffany inherited a germline mutation in a tumor suppressor gene on chromosome 22 that provided susceptibility. When somatic mutations occurred in the homologous gene in Schwann cells along the eighth cranial nerve, the tumors grew. Tiffany had successful surgery.

Concerned that other family members might be affected and knowing that surgery is only effective if the tumors are caught early, Tiffany urged her parents to be checked even though they had no symptoms. She knew one of them had to be affected because her illness was inherited as an autosomal dominant trait. Her father Bill was unaffected, but her mother Monica had two very small tumors. Tiffany's brother Tony had had tinnitus for years, and when he was checked, he, too, had a small tumor. Neither Tony nor Tiffany are married or have children. Tiffany finds out that her maternal grandmother, Lila, died at age 72 from progressive weakness, paralysis, and eventually respiratory failure—symptoms consistent with advanced vestibular Schwannoma.

Worksheet

1. Draw a pedigree of Tiffany's family.

2. Someone with sporadic vestibular Schwannoma cannot pass the disorder onto the next generation because

_____

_____.

3. A consensus statement on vestibular Schwannoma by the National Institutes of Health reads, in part, "Thus the trait is recessive at the cellular level, but exhibits a dominant pattern of genetic transmission in families." What does this mean?

4. Two explanations for a person with vestibular Schwannoma tumors in both ears forming grapelike masses, but who does not have affected parents are:

   a. _____

   b. _____

5. Vestibular Schwannoma tends to be more severe (earlier onset, more abundant and faster-growing tumors) if it is inherited from one's mother. This is an example of

   a. incomplete penetrance.
   b. pleiotropy.
   c. mutagenesis.
   d. genomic imprinting.
   e. genetic engineering.

VON HIPPEL-LINDAU DISEASE

Key Words

Concordance	Pedigree
Genetic marker	Penetrance
Multifactorial inheritance	RFLP
Oncogene	Tumor suppressor

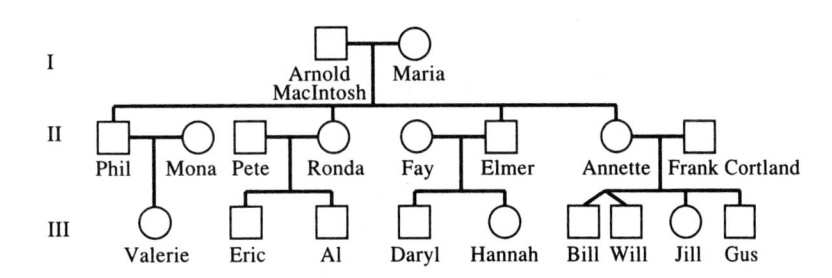

```
I                          ┌──┬──○
                          Arnold Maria
                        MacIntosh

II    ┌──┬──○   ┌──┬──○   ○──┬──┐   ○──┬──┐
      Phil Mona Pete Ronda Fay Elmer Annette Frank Cortland

III      ○       ┌──┐      ┌──○      ┌─┬─┐  ○  ┌
      Valerie  Eric  Al   Daryl Hannah Bill Will Jill Gus
```

The MacIntosh and Cortland families have several members who have tumors. Arnold died of tumors in both kidneys. His son Elmer developed a brain tumor as a child that was successfully removed, and he then developed an eye tumor in young adulthood that was also successfully removed.

Annette, the mother of four young children, died at age 42 of a pancreatic tumor. Her doctor suspects that the families may have von Hippel-Lindau disease, an autosomal dominant condition that predisposes individuals to develop tumors of the retina, central nervous system (brain and spinal cord), kidneys, adrenal glands, pancreas, and epididymis (a male reproductive structure).

A geneticist confirms the diagnosis and advises the youngest generation to be tested with magnetic resonance imaging to spot brain and spinal cord tumors and computerized tomography scans to detect kidney, adrenal, and pancreatic tumors. Fay and Elmer decide to have their children tested. A small tumor is found in Daryl's kidney. His surgery is a success. Hannah is free of tumors. Frank allows Jill and Gus to be checked because they are teens and asked to be tested, but did not give permission for Bill and Will to be examined because they are very young. Jill has a small, operable, brain tumor.

Undergoing frequent scans for tumors is stressful and costly. As an alternative, the geneticist enrolls the family in a research project in which several restriction fragment length polymorphisms are traced in each family member. If RFLPs can be found that are only in those who have tumors, then the presence of those markers in healthy relatives under the age of typical onset provides a presymptomatic diagnosis.

RFLP markers designated A, B, C, and D are found near the suspected site of the von Hippel-Lindau gene on chromosome 3p. Arnold, Elmer, and Annette, who had tumors, have markers A and C, as do Daryl and Jill, whose symptoms were revealed with scans. Phil, Ronda, Hannah, and Gus have RFLPs C and D.

Worksheet

1. Fill in affected individuals in the pedigree.

2. If Bill has the gene for von Hippel-Lindau disease, the chance that Will does too is _____.

3. A genetic marker test is more definitive than checking repeatedly for symptoms because _____.

4. Mona, Pete, Fay, and Frank are tested and found to have RFLPs B and D. Why is this information important in making presymptomatic diagnoses on their children?

125

5. Von Hippel-Lindau disease is incompletely penetrant, and penetrance is age dependent. By age 34, 78% of those inheriting the gene develop tumors. If, by age 34, Bill has a tumor but Will doesn't, the difference in gene expression could be attributed to _____.

6. If the mutant allele behind this disorder is a deletion, would the wild type allele more likely be an oncogene or a tumor suppressor? Cite a reason for your answer.

VON WILLEBRAND DISEASE

Key Words

Exon	Kilobase
Genetic code	Mutation
Genetic heterogeneity	Phenotype
Intron	Restriction mapping

Von Willebrand disease is the most common inherited bleeding disorder, and it is transmitted as an autosomal recessive trait. Eighty percent of sufferers have a mild form of the illness, with prolonged bleeding from a wound. About 1% of sufferers have a very severe form. The disorder is treated with clotting factors concentrated from blood donors.

A mutation in the gene encoding von Willebrand factor (vWF) causes von Willebrand disease. This is a large, multi-subunit plasma glycoprotein that interacts with platelets and clotting factor VIII (the protein absent in hemophilia A) to control blood clotting. People with severe von Willebrand disease lack vWF.

The vWF gene maps to chromosome 12p. It spans 178 kilobases, includes 52 exons, and is transcribed into an mRNA of 8.8 kilobases. The gene product is translated into a precursor form with only the mature protein forming vWF as follows:

signal peptide	propeptide	mature protein
22 amino acids	741 amino acids	2050 amino acids

Swedish researchers sequenced the vWF gene from twenty-five patients with the severe form of the illness. All had a point mutation altering a CGA mRNA codon to a UGA codon. One patient had this particular mutation in an intron, where it introduces a new recognition site for the restriction enzyme DdeI.

Worksheet

1. The number of DNA bases that are not represented in the mature protein product of the vWF gene is _____ .

2. The number of DNA bases encoding the form of vWF found in the bloodstream is _____ .

3. The point mutation identified in the twenty-five von Willebrand disease patients is a _____ and _____ mutation.

    a. transversion, nonsense
    b. transversion, missense
    c. transition, nonsense
    d. transition, missense

4. The dinucleotide CG is a mutational hot spot, meaning that it is implicated in mutations causing a variety of disorders. Why might this dinucleotide be particularly prone to mutations that drastically alter the phenotype?

5. A mutation that adds a restriction site for DdeI results in pieces of DNA, following a restriction digest with the enzyme, that are larger or smaller than for the wild type allele?

6. There is also a sex-linked form of von Willebrand disease. How does the structure and function of vWF explain this genetic heterogeneity?

7. The phenotypes of von Willebrand disease and hemophilia A are very similar. What information could be used to distinguish between the two disorders?

The Genetic Code

First Letter	Second Letter				Third Letter
	U	C	A	G	
U	UUU ⎫ phenylalanine UUC ⎭ (phe) UUA ⎫ leucine (leu) UUG ⎭	UCU ⎫ UCC ⎬ serine (ser) UCA ⎪ UCG ⎭	UAU ⎱ tyrosine (tyr) UAC ⎰ UAA STOP UAG STOP	UGU ⎱ cysteine (cys) UGC ⎰ UGA STOP UGG tryptophan (try)	U C A G
C	CUU ⎫ CUC ⎬ leucine (leu) CUA ⎪ CUG ⎭	CCU ⎫ CCC ⎬ proline (pro) CCA ⎪ CCG ⎭	CAU ⎱ histidine (his) CAC ⎰ CAA ⎱ glutamine (gln) CAG ⎰	CGU ⎫ CGC ⎬ arginine (arg) CGA ⎪ CGG ⎭	U C A G
A	AUU ⎫ AUC ⎬ isoleucine (ilu) AUA ⎭ AUG+ methionine (met)	ACU ⎫ ACC ⎬ threonine (thr) ACA ⎪ ACG ⎭	AAU ⎱ asparagine (asn) AAC ⎰ AAA ⎱ lysine (lys) AAG ⎰	AGU ⎱ serine (ser) AGC ⎰ AGA ⎱ arginine (arg) AGG ⎰	U C A G
G	GUU ⎫ GUC ⎬ valine (val) GUA ⎪ GUG ⎭	GCU ⎫ GCC ⎬ alanine (ala) GCA ⎪ GCG ⎭	GAU ⎱ aspartic acid (asp) GAC ⎰ GAA ⎱ glutamic acid (glu) GAG ⎰	GGU ⎫ GGC ⎬ glycine (gly) GGA ⎪ GGG ⎭	U C A G

APPENDIX A

Symbols:

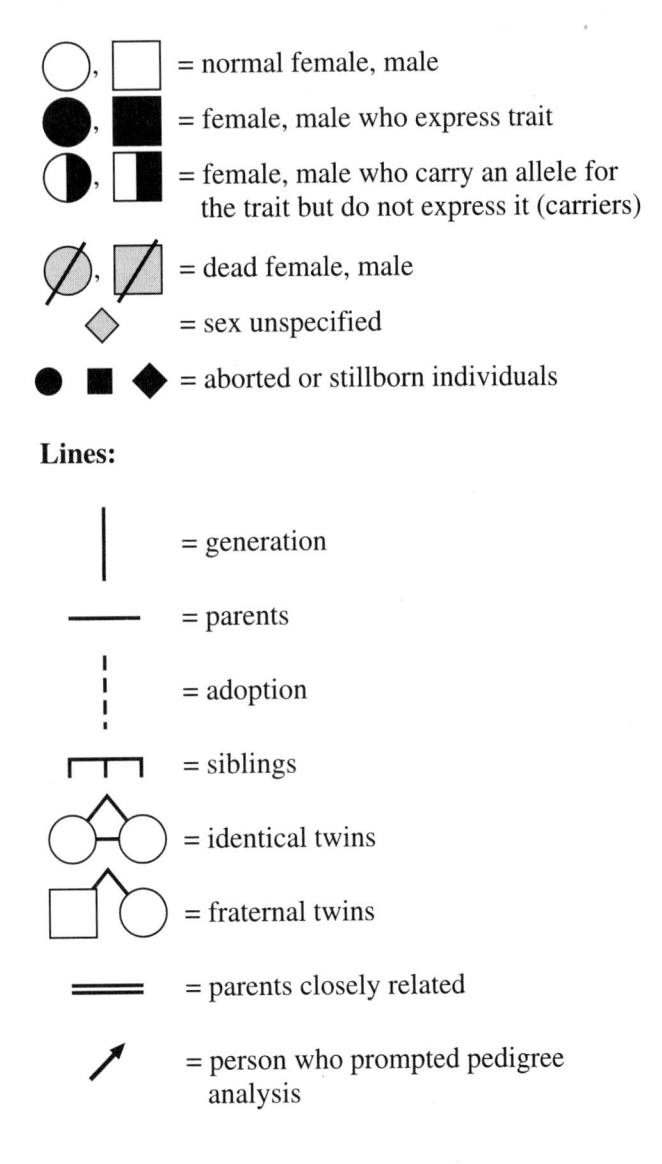

◯, ▢ = normal female, male

●, ■ = female, male who express trait

◐, ◨ = female, male who carry an allele for the trait but do not express it (carriers)

⌀, ⌀ = dead female, male

◆ = sex unspecified

● ■ ◆ = aborted or stillborn individuals

Lines:

| = generation

— = parents

┊ = adoption

⊓ = siblings

◯–◯ (with apex) = identical twins

▢ ◯ (with apex) = fraternal twins

= = parents closely related

↗ = person who prompted pedigree analysis

Numbers:

Roman numerals = generations

Arabic numerals = individuals

Symbols used in pedigree construction are connected to form a pedigree chart, which displays the inheritance patterns of particular traits.

HUMAN GENETICS GLOSSARY

Acrocentric A chromosome with the centromere located close to one tip.

Acrosome A protrusion on the anterior end of a sperm cell containing digestive enzymes that enable the sperm to penetrate the protective layers around the oocyte.

Active immunity An immune response mounted by the individual's immune system, rather than from a parent or other source of immune system cells or biochemicals.

Adenine One of two purine nitrogenous bases in DNA and RNA.

Adenosine deaminase The enzyme deficient in severe combined immune deficiency, in which both humoral and cellular immunity are impaired because T cells and then B cells are destroyed.

Alike in state Inheriting identical alleles from parents who are not blood relatives to each other.

Allantois An extra-embryonic membrane that manufactures blood cells and gives rise to the blood vessels of the umbilical cord.

Allele An alternate form of a gene.

Alpha-1-antitrypsin A glycoprotein normally present in blood serum that helps microscopic air sacs in the lungs inflate and function properly. Lack of the enzyme results in hereditary emphysema.

Alpha fetoprotein A protein that enters the maternal circulation from the fetal liver at a known rate throughout pregnancy. Elevated AFP in a pregnant woman's blood can indicate a neural tube defect in the fetus.

Amino acid A small organic molecule that is a building block of a protein. Contiguous triplets of DNA nucleotide bases encode the twenty types of amino acids that are polymerized to form biological proteins.

Amniocentesis A prenatal diagnostic procedure in which a needle is inserted into the uterus to remove a small sample of amniotic fluid, which contains fetal cells and biochemicals. A chromosome chart is constructed from cultured fetal cells, and tests for certain inborn errors of metabolism conducted on fetal biochemicals. Amniocentesis is performed during the 16th week of pregnancy.

Amniotic cavity A space between an early embryo and the uterine lining.

Anaphase The stage of mitosis when centromeres split and the two sets of chromosomes move to opposite ends of the cell. In anaphase of meiosis I, homologs separate.

Aneuploid A cell with one or more extra or missing chromosomes.

Angiotensinogen A protein elevated in the blood of people with high blood pressure.

Ankyrins Proteins that hold together cell membranes.

Antibody A multisubunit protein produced by B cells that latches onto a specific nonself antigen at one end, alerting other components of the immune system or directly destroying the antigen.

Anticodon A 3-base sequence on one loop of a transfer RNA molecule that is complementary to an mRNA codon and, therefore, brings together the appropriate amino acid and its mRNA instructions.

Antigen binding site The specialized end of an antibody molecule that binds antigen.

Antigen processing The process by which a viral protein infecting a macrophage is cut into pieces and transported through the cell's inner membrane network. Here it binds to an HLA protein, which transports it to the cell surface where it is displayed, alerting the immune system.

Antiparallel The head-to-tail arrangement of the two entwined chains of the DNA double helix.

Antisense RNA RNA complementary to RNA that encodes protein.

Antisense strand The strand of the DNA double helix for a particular gene that is not transcribed into mRNA.

Antisense technology Use of nucleic acids with sequences complementary to genes or protein-encoding RNA to silence gene function.

Apolipoproteins Protein portions of lipoproteins, which contain and transport fat (lipid).

Artificial insemination The technique of placing donated sperm in a woman's reproductive tract to achieve fertilization when the woman's mate is infertile.

Artificial selection Selective breeding. Choosing particular individuals to reproduce, based on the perceived value of their inherited characteristics.

Autoantibodies Antibodies that attack self antigens, causing an autoimmune disorder in which the body damages its own tissues.

Autoimmunity Production of autoantibodies that attack the body's own tissues. There are several autoimmune disorders, many of them characterized by inflammation of the affected tissues.

Autosomal dominant A trait affecting either sex caused by an allele on an autosome that exerts an observable effect when present in one copy.

Autosomal recessive A trait affecting either sex caused by an allele on an autosome that must be present in two copies to exert an observable effect.

Autosome A non-sex chromosome.

B cells White blood cells (lymphocytes) that secrete antibody proteins in response to recognizing nonself molecules.

Bacteriophage A virus that infects bacteria. Bacteriophages are used as vectors to transfer DNA from one cell to another in recombinant DNA technology.

Balanced polymorphism Maintenance of a harmful recessive allele in a population because the heterozygote enjoys a survival or reproductive advantage.

Barr body The dark-staining body seen in the nucleus of a cell from a female mammal, corresponding to the inactivated X chromosome.

Basement membranes Boundaries between tissues. Cancer cells often anchor here and secrete substances that ease their spread from one tissue section to another.

Beta amyloid A protein that accumulates in the brains of people who have Alzheimer's disease. Amyloid-like proteins also build up in a class of genetic disorders called amyloidoses.

Bioremediation Using microorganisms, possibly altered by genetic engineering, to detoxify environmental pollutants.

Biotechnology The alteration of cells or biological molecules with a specific application including monoclonal antibody technology, genetic engineering, or cell culture.

Blastocyst The preembryonic stage of human development when the organism is a hollow, fluid-filled ball of cells.

Blastomere A cell in a preembryonic organism resulting from cleavage divisions.

Body mass index A measure of weight taking height into account (weight/height^2). This is a multifactorial trait.

Callus A lump of undifferentiated plant somatic tissue growing in culture.

Cancer A group of disorders resulting from the loss of normal control over mitotic rate and number of divisions.

Candidate gene A DNA sequence that may be a sought after gene. Once a candidate gene is identified, a series of tests is conducted to show its involvement in a particular disorder or trait.

Cap A short sequence of modified nucleotides at the start of an mRNA molecule.

Capacitation Activation of sperm cells in the human female reproductive tract.

Carcinogen A substance that induces cancerous changes in a cell.

Carrier An individual who is heterozygous for a particular gene.

Cell The structural and functional unit of life.

Cell cycle The life of a cell in terms of whether it is dividing or is in interphase.

Cell membrane (plasmalemma) An oily structure built of proteins embedded in a lipid bilayer that forms the boundary of cells.

Cellular immune response The part of the immune response provided by T cells that travel to where they are needed and release cytokines.

CentiMorgan A unit of gene mapping equal to about one million DNA bases.

Central dogma of molecular biology A model depicting the flow of genetic information from DNA to RNA to protein. This is correct but is an oversimplification of the biological interaction of nucleic acids and proteins.

Centrioles Paired, oblong structures built of microtubules and found in animal cells where they organize the mitotic spindle.

Centromere A characteristically-located constriction in a chromosome.

Cervix The opening to the uterus in the female human.

Chaperone proteins Proteins that stabilize partially folded portions of a polypeptide chain as it forms.

Chlorophyll A green pigment used by plants to harness the energy in sunlight.

Chloroplast A plant cell organelle housing the reactions of photosynthesis.

Chord distance A physical depiction of the degree of relationship of populations based on allele frequencies.

Chorionic villi Fingerlike projections extending from the chorion (an extraembryonic membrane) to the uterine lining.

Chorionic villus sampling A technique of prenatal diagnosis that is based on analyzing chromosomes in chorionic villus cells that, like the fetus, descend from the fertilized ovum.

Chromatid A continuous strand of DNA comprising an unreplicated chromosome or one half of a replicated chromosome.

Chromatin DNA and its associated histone proteins.

Chromosome A dark-staining, rod-shaped structure in the nucleus of a eukaryotic cell built of a continuous molecule of DNA, wrapped in protein.

Chromosome mosaic An individual in whom some cells have a particular chromosomal anomaly, and others do not.

Chromosome walk A technique of overlapping short pieces of a chromosome to approach a region about which little is known.

Cleavage A period of rapid cell division following fertilization but before embryogenesis.

Clines Allele frequencies that differ greatly between different populations.

Cluster-of-differentiation (CD4) antigen A surface protein on certain T cells that begin an immune response by recognizing a macrophage presenting a nonself antigen. T cells bearing CD4 antigens are early targets of the virus thought to cause AIDS.

Codominant Alleles that are both expressed in the heterozygote.

Codon A continuous triplet of mRNA that specifies a particular amino acid.

Coefficient of relationship A value that indicates the closeness of genetic relationship (percentage of genes shared) between two individuals.

Colony stimulating factors A class of cytokines that stimulate bone marrow to produce lymphocytes.

Colostrum Milk produced in the first few days following birth. It is rich in antibodies.

Complementary The tendency of adenine to hydrogen bond to thymine and guanine to cytosine in the DNA double helix.

Complementary DNA (cDNA) A DNA molecule that is reverse transcribed from mRNA, thereby, reflecting gene activity in a cell.

Completely dominant An allele whose presence totally masks the expression of another recessive allele.

Completely penetrant A disease-causing allele combination that is always expressed in individuals who have it.

Concordance A measure of the inherited component of a trait, consisting of the number of pairs of either monozygotic or dizygotic twins in which both members express a trait, divided by the number of pairs in which at least one twin expresses the trait.

Conformation The three-dimensional shape of a molecule, such as a protein.

Consanguinity Being blood relatives; sharing a common ancestor.

Conservative DNA replication An incorrect model of DNA replication in which an entire new double helix is made from the parental double helix. It was ruled out when Meselson and Stahl used density shift experiments to demonstrate semiconservative replication.

Constant region The lower portion of an antibody amino acid chain that tends to be very similar in different species.

Constitutional mutation A mutation that occurs in every cell in an individual, indicating that it arose in the germline of one of the parents.

Contact inhibition The tendency of a cell to cease dividing once it touches another cell.

Contig map A chromosome map constructed by overlapping small maps.

Continuously variable A trait that varies greatly among individuals, such as height and intelligence.

Corona radiata Cells surrounding the secondary oocyte and the zona pellucida.

Coupling For two linked genes (each heterozygous in an individual), the condition where each chromosome carries a recessive allele of one gene and a dominant allele of the other.

CpG islands Regions of the genome containing many repeats of the sequence CGCGCG. Genes encoding proteins are often close to such sequences.

Critical period The time during prenatal development when a specific structure can be altered by a gene or an external influence.

Crossing over The exchange of genetic material between homologous chromosomes during prophase of meiosis I.

Cyclin A protein controlling the cell cycle.

Cystic fibrosis transmembrane conductance regulator The chloride channel protein that is abnormal in cystic fibrosis.

Cytochrome C A protein involved in cellular respiration that is remarkably similar in sequence among many species, indicating that it is ancient and essential. It has been highly conserved in evolution.

Cytogenetic map A depiction of a chromosome including bands produced by staining, also including indications of where known chromosome abnormalities reveal the locations of particular genes.

Cytogenetics Matching phenotypes to detectable chromosomal abnormalities.

Cytokines Immune system biochemicals released from T cells.

Cytoplasm The jellylike fluid in which organelles are suspended in eukaryotic cells.

Cytosine One of the two pyrimidine nitrogenous bases in DNA and RNA.

Cytoskeleton A framework built of arrays of protein rods and tubules found in animal cells.

Cytotoxic T cells White blood cells that attack nonself cells by attaching to them and releasing chemicals.

Daughter DNA strand The most recently synthesized half of a DNA double helix.

Dedifferentiated A cell that is less specialized than the cell from which it descends. A characteristic of cancer cells.

Degenerate Different codons specifying the same amino acid.

Deletion A missing sequence of DNA or part of a chromosome.

Density shift experiment Shifting bacterial cultures labelled with radioactive phosphorus or sulfur between temperatures to tell whether the biochemical that replicates is DNA or protein.

Deoxyribonucleic acid (DNA) A double-stranded nucleic acid built of nucleotides containing a phosphate group, a nitrogenous base (A,T,G or C) and the sugar deoxyribose.

Deoxyribose The 5 carbon sugar in a DNA nucleotide.

Dicot A plant with two seed leaves that appear as the plantlet emerges from the soil.

Differential gene expression Activation and suppression of subsets of genes that sculpts the characteristics of specific cell types.

Dihybrid An individual heterozygous for two particular genes.

Diploid A cell with two copies of each chromosome.

Dispersive DNA replication Erroneous model of DNA replication in which the DNA double helix falls apart and picks up new nucleotides to form two new double helices.

Dizygotic twins Fraternal twins.

DNA fingerprinting A method used to distinguish individuals based on differences in DNA sequence.

DNA hybridization A technique using the extent of complementary base pairing to estimate how similar the genomes of two species are.

DNA library A collection of recombinant bacterial cultures containing small pieces of an entire genome of another organism.

DNA polymerase A type of enzyme that participates in DNA replication by inserting new bases and correcting mismatched base pairs.

DNA probe A short sequence of DNA corresponding to a specific gene under investigation that is labelled with a radioactive isotope. When the probe is applied to a biological sample, its complementary base pairs with its corresponding sequence whose locus is revealed by the radioactive signal.

DNA replication Construction of a new DNA double helix using the information in parental strands as a template.

Dominant An allele that masks the expression of another allele.

Duplication An extra copy of a gene or DNA sequence, usually caused by misaligned pairing in meiosis.

Ectoderm The outermost embryonic germ layer whose cells become part of the nervous system, sense organs, and outer skin layer and its specializations.

Ectopic pregnancy The implantation of a zygote in the wall of a Fallopian tube, rather than in the uterus. The zygote must be removed to save the woman's life.

Elastase An enzyme whose levels are controlled by alpha-1-antitrypsin. When alpha-1-antitrypsin is deficient, elastase levels rise, and lung tissue is destroyed. This is hereditary emphysema.

Electroporation A technique in which a brief jolt of electricity is used to open transient holes in cell membranes allowing foreign DNA to enter.

Elongation The stage of protein synthesis when the ribosome binds to the initiation complex and amino acids are joined.

Embryo The stage of prenatal development when organs develop from a three-layered organization.

Embryo adoption A reproductive technology in which a woman is artificially inseminated and conceives. A week later, the preembryo is flushed out of her uterus, and transferred to the uterus of the wife of the man who donated the sperm.

Embryonic induction The ability of a group of specialized cells in an embryo to stimulate neighboring cells to specialize.

Embryonic stem cell A cell in a mammalian embryo that has not yet specialized. Embryonic stem cells are used in gene targeting.

Empiric risk Predicting the likelihood of an individual developing a disorder based on observed frequencies of the disorder among similar individuals.

Endoderm The innermost embryonic germ layer whose cells become the organs and linings of the digestive, respiratory, and urinary systems.

Endometriosis A condition that causes tissue buildup in and on the uterus, monthly bleeding, and severe cramps.

Endometrium The inner uterine lining.

Endothelium The tilelike, single-celled inner lining of blood vessels.

Enzyme A protein that catalyzes a specific type of chemical reaction.

Epidermal growth factor A growth factor that stimulates cell division to form or repair linings.

Epistasis A gene whose function masks another gene's expression.

Equational division The second meiotic division when four haploid cells are generated from two haploid cells, which are products of meiosis I, by a mitosis-like division.

Equilibrium density gradient centrifugation The technique used by Meselson and Stahl to distinguish newly replicated DNA from parental DNA.

Eugenics The control of individual reproductive choices in order to achieve a societal goal.

Eukaryotic cell A complex cell containing organelles that carry out a variety of specific functions.

Euploid A somatic cell with the correct number of chromosomes for that species. The human euploid chromosome number is 23 pairs.

Evolutionary conservation Similarity in sequence of a gene or protein between two species.

Excision repair The enzyme-catalyzed removal of pyrimidine dimers in DNA that corrects errors in DNA replication.

Exon The DNA base sequences of a gene that encode amino acids. Exons are interspersed with noncoding regions called introns.

Explant A small piece of plant tissue used to start a culture.

Expressivity The degree of expression of a phenotype.

Extraembryonic membranes Structures that support and nourish the mammalian embryo and fetus, including the yolk sac, allantois, and amnion.

Fallopian tubes In the human female, paired tubes leading from near the ovaries to the uterus where oocytes can be fertilized.

Fertilized ovum An oocyte after it has been penetrated by a sperm.

Fibrillin An elastic connective tissue protein that is abnormal in Marfan syndrome.

Fibroblast A cell of connective tissue that secretes the proteins collagen and elastin.

Fibroid tumors Benign tumors occurring in and on the uterus.

Filial generation A second generation.

Follicle cells Cells surrounding oocytes that provide nourishment and protection.

Foreign antigen A molecule that stimulates B cells to manufacture antibodies. The immune system perceives foreign antigens as nonself.

Founder effect A type of genetic drift in human populations where a few members leave to found a new settlement taking with them a subset of the alleles in the original population.

Free radicals Highly reactive by-products of metabolism that can damage tissue.

Freemartin A pair of male-female cattle twins in whom the male is fertile but the female is infertile.

Fusion protein A protein resulting from one gene moving next to another. Both genes are transcribed and translated as if they encode a single gene product.

G1 phase The stage of interphase when proteins, lipids, and carbohydrates are synthesized.

G2 phase The stage of interphase when membrane components are synthesized and stored.

Gamete A sex cell. The sperm and ovum are sex cells.

Gamete intrafallopian transfer (GIFT) A reproductive technology in which sperm are transferred from the laboratory to a woman's Fallopian tube where fertilization occurs.

Ganciclovir A drug that destroys cells that manufacture an enzyme called thymidine kinase encoded by a herpes virus. Some genetic engineering experiments endow cells with this viral gene so that their destruction can be controlled.

Gastrula The point in prenatal development when cells are no longer in the same position relative to each other. The point when infoldings begin to form.

Gene A sequence of DNA that specifies the sequence of amino acids in a particular polypeptide.

Gene amplification A technique that uses enzymes of DNA replication in vitro to make many copies of a particular DNA sequence.

Gene flow Movement of alleles between populations.

Gene library The genome of an organism is cut into pieces that are each cultured in recombinant bacteria.

Gene mapping Locating a gene's position on a chromosome.

Gene pool All the genes in a population.

Gene targeting A form of genetic engineering in which an introduced gene exchanges places with its counterpart on a host cell's chromosome by homologous recombination.

Gene therapy Replacing a malfunctioning gene.

Genetic code The correspondence between specific DNA base sequences and the amino acids that they specify.

Genetic drift Change in gene frequencies when small groups of individuals are separated from or leave a larger population.

Genetic engineering Manipulation of genetic material, including: altering DNA in an organism to suppress or enhance its activity, or combining genetic material from different species.

Genetic heterogeneity Different genotypes that have identical phenotypes. When the same symptoms reflect disorders with different modes of inheritance.

Genetic load The collection of deleterious alleles in a population.

Genetic map A depiction of the relative locations of genes along a chromosome based on the frequency of crossing over between them.

Genetic marker A detectable piece of DNA that is closely linked to a gene of interest whose precise location is not known.

Genome All of the DNA in a cell of an organism.

Genomic imprinting A phenotype that is different depending upon the sex of the parent passing on the causative gene or chromosome abnormality.

Genotype The genetic constitution of an individual.

Genotypic ratio The ratio of genotypes resulting from a particular cross.

Germline cells Diploid cells in the gonads from which gametes form by meiotic cell division.

Germline mutation A mutation that occurs in every cell in an individual and, therefore, was inherited from a parent.

Glia Cells that support nerve cells.

Glucocerebrosidase An enzyme that is missing in the autosomal recessive inborn error of metabolism Gaucher disease, causing swelling of the liver and spleen and early death.

Glycoprotein A molecule built of a protein and a sugar.

Growth factor A locally-acting protein that assists in wound healing.

Guanine One of the two purine nitrogenous bases in DNA and RNA.

Haploid A cell with one copy of each chromosome.

Haplotype A set of specific linked alleles in an individual.

Hardy Weinberg equilibrium An idealized state in which gene frequencies in a population do not change between generations.

Hayflick limit A cell's internal clock controlling the number of cell divisions.

Heavy chain Two of the four longer amino acid chains of an antibody molecule subunit.

Helper T cells White blood cells (lymphocytes) that activate various immune system functions, including stimulating B cells to produce antibodies, activating cytotoxic T cells, and secreting cytokines.

Hemizygous A gene carried on the Y chromosome in humans.

Heritability (1) An estimate of the proportion of variation in phenotype among a group of individuals that can be attributed to heredity. (2) The fact that a cell descended from a cancer cell is also cancerous.

Heritable (germline) gene therapy Altering a gene in a gamete or fertilized ovum, so the change is perpetuated in all cells of the resulting individual.

Herpes simplex virus A virus that causes cold sores and genital lesions. The virus is also used to carry genes into nerve cells in an attempt to treat neurological disorders by gene therapy.

Heterochromatin Dark-staining genetic material that is believed to be inactive.

Heterogametic sex The sex with two different sex chromosomes, such as the human male.

Heterozygote An individual who has two different alleles for a particular gene.

High resolution chromosome banding Staining of chromosomes in early mitosis, revealing many bands.

Histone A protein around which DNA entwines.

Homeobox A 60 base sequence of DNA that is found in genes in many organisms that control differentiation of tissues in different segments of the organism. Homeobox genes are involved in making developmental decisions.

Hominid Animals ancestral to humans only.

Hominoid Animals ancestral to apes and humans only. They lived 22 to 32 million years ago in Africa.

Homogametic sex The sex with two identical sex chromosomes, such as the human female.

Homologous chromosomes Chromosomes that have the same sequence of genes.

Homologous recombination A natural process by which a piece of DNA detects and displaces an identical or very similar DNA sequence in a chromosome.

Homozygote An individual who has two identical alleles for a particular gene.

Homozygous dominant When an individual has two identical dominant alleles of a particular gene.

Homozygous recessive When an individual has two identical recessive alleles of a particular gene.

Hormone A biochemical manufactured in a gland and transported in the blood to a target organ where it exerts a characteristic effect.

Human chorionic gonadotropin A hormone secreted by the preembryo and embryo that prevents menstruation.

Human immunodeficiency virus (HIV) A retrovirus implicated in causing AIDS. It binds to and enters CD4-bearing helper T cells, preventing them from further activating immunity, and using them to reproduce.

Human leukocyte antigen (HLA) complex A closely linked set of genes on the short arm of chromosome 6 that encode cell surface proteins.

Humoral immune response The part of the immune response provided by antibodies secreted into the bloodstream by B cells.

Hybridize Binding of a sequence of DNA to its complementary sequence.

Hybridoma An artificial cell that produces monoclonal antibodies (large amounts of a single antibody type), created by fusing a cancer cell with a B cell.

Identical by descent Identical alleles in an individual inherited from parents who are blood relatives and have the allele in common because of a shared ancestor.

Ideogram A diagram of a chromosome showing bands and locations of known genes. An ideogram combines cytogenetic and molecular information.

Idiotype The specific part of the antigen binding site of an antibody that fits itself around a particular foreign antigen.

Immunogenetics Study of the genes that provide immunity.

Immunotherapy Using cells and biochemicals of the immune system to fight disease, particularly cancer.

Implantation Nestling of the blastocyst into the uterine lining.

In situ hybridization Binding of a DNA probe to its complementary sequence in a DNA sample.

In vitro fertilization Fertilization of an oocyte by a sperm in a piece of laboratory glassware.

Inborn error of metabolism A disorder caused by a missing or inactive enzyme. The symptoms are caused by buildup of the substrate and deficit of the biochemical into which the substrate is metabolized.

Inbreeding Mating between blood relatives that produces offspring.

Incomplete dominance When the phenotype of a heterozygote is intermediate between the phenotypes of the two homozygotes.

Incompletely penetrant An allele combination that is not expressed in all individuals who inherit it.

Independent assortment The random arrangement of maternally and paternally derived homologs during metaphase of meiosis I.

Infertility The inability to conceive a child after a year of unprotected intercourse.

Informative An RFLP (genetic marker) that is always present in individuals who have inherited the marked disorder, but never in those who have not inherited it, within a specific family.

Initiation The start of protein synthesis, when mRNA, tRNA carrying an amino acid, ribosomes, energy storing molecules, and protein factors begin to assemble.

Initiation complex The aggregation of the components of the protein synthetic apparatus just before mRNA is translated into a polypeptide chain.

Initiation site The site where DNA replication begins on a chromosome.

Inner cell mass The cells in the blastocyst that develop into the embryo.

Interferon An immune system biochemical released by T cells. Interferon fights viral infections and is used as a drug to treat a certain type of leukemia.

Interleukins A class of cytokines that respond to viral infections and are used as drugs to treat certain types of cancers.

Interphase The period of the cell cycle when the cell synthesizes proteins, lipids, carbohydrates, and nucleic acids. Interphase is the period of the cell cycle when the cell is not actively dividing.

Intron Base sequences within a gene that are transcribed but are excised from the mRNA before translation into protein. Introns are interspersed with protein-encoding exons.

Invasiveness The ability of cancer cells to squeeze into tight places.

Inversion A chromosome with part of its gene sequence inverted.

Isochromosome A chromosome with identical arms that form when the centromere splits in the wrong plane.

Karyotype A size-order chart of chromosomes.

Kuru A neurological degeneration caused by infection by a slow virus obtained from uncooked human flesh.

L gene A blood group with three co-dominant alleles, M, N, and S.

Leader sequence A short sequence at the start of mRNA that enables it to bind to rRNA in a ribosome.

Lethal allele A recessive allele that, when homozygous, causes death of the organism.

Lewis A gene that encodes an enzyme that adds an antigen to the sugar fucose, which is placed on red blood cells by the product of the H gene.

Ligase An enzyme that catalyzes the formation of covalent bonds in the sugar-phosphate backbone of DNA.

Light chain The two shorter of the four polypeptide chains comprising an antibody molecule subunit.

Linkage The location of genes on the same chromosome.

Linkage disequilibrium Extremely tight linkage between two genes that are very close to one another on a chromosome.

Linkage map A diagram of the relative positions of genes on chromosomes.

Lipid bilayer A two-layered structure formed by the alignment of phospholipids, reflecting their chemical attractions and repulsions to water.

Lipoprotein A large molecule built of a protein and a lipid. Lipoproteins transport cholesterol in the bloodstream.

Lipoprotein lipase A protein that breaks down fats on the inside walls of small blood vessels.

Liposome A fatty bubble used to carry molecules, including DNA.

Logarithm of the odds (LOD) A calculation that expresses how close two genes are to each other on a chromosome. An LOD score indicates how likely it is that the two genes are inherited together by chance.

Lymphocyte A type of white blood cell that is made in the bone marrow and migrates to the lymph nodes, spleen, tonsils, and thymus gland; it circulates in the blood and tissue fluid. Lymphocytes include the B and T cells that provide immunity.

Lysosomal storage disease A class of inborn errors of metabolism in which a lysosomal enzyme is absent or abnormal, resulting in buildup of its substrate.

Lysosome A sac in a eukaryotic cell in which molecules and worn out organelles are enzymatically dismantled.

Macroevolution Genetic change sufficient to form a new species.

Macrophage A wandering scavenger cell that assists in providing immunity.

Malignant tumor A cancerous tumor; a tumor that grows and infiltrates surrounding tissue.

Manifesting heterozygote A female carrier of a sex-linked recessive trait who expresses the phenotype because the X chromosome with the wild type allele is inactivated in affected tissues.

Map unit An estimate of the physical distance between two linked genes based on how frequently crossovers occur between them.

Maternal inheritance A trait passed from the mother and encoded by a mitochondrial gene. Sperm do not usually contain mitochondria, so males do not transmit mitochondrial genes.

Maturation The period following meiosis when the distinctive characteristics of sperm and egg form.

Maturation-promoting factor A protein that controls the cell cycle.

Meiosis Cell division that halves the genetic material.

Membrane A biological structure built of proteins embedded in and protruding from a lipid bilayer that surrounds cells or compartmentalizes structures within.

Memory cells Cells that develop from activated B cells and enable the immune system to remember a nonself antigen that has been encountered before.

Mendelian trait A trait specified by a single gene.

Mesoderm The middle embryonic germ layer whose cells become bone, muscle, blood, dermis, and reproductive organs.

Messenger RNA A molecule of ribonucleic acid that is complementary in sequence to the sense strand of a gene.

Metacentric A chromosome with the centromere located approximately in the center.

Metaphase The second stage of cell division when chromosomes align down the center of a cell. In mitosis the chromosomes form a single line. In meiosis I the chromosomes line up in homologous pairs.

Metastasis The spreading of cancer from its site of origin to other parts of the body.

Microevolution The result of allele frequency changes in a population.

Microfilaments Solid protein rods beneath cell membranes that contract just before a cell divides, helping to separate the daughter cells.

Microtubules Long, hollow tubules built of the protein tubulin that provide movement within cells.

Mitochondrion The organelle along whose internal membranes the reactions of cellular metabolism occur.

Mitosis A form of cell division in which two genetically identical cells are generated from one. The genetic material is replicated and organelles and other subcellular structures apportioned between the two daughter cells.

Molecular clock An estimate of the time since the divergence of two species from a shared ancestor. The estimate is based upon the differences in the sequences of genes or proteins of two species, assuming a known and a constant rate of mutation.

Molecular evolution Comparing protein and gene sequences in order to determine how recently closely related species diverged from a common ancestor.

Molecular genetics The study of the structure and function of DNA and its relationship to RNA and proteins.

Molecular systematics A branch of molecular evolution in which comparing gene sequences reveals genetic change that cannot yet be observed in physical characteristics.

Molecular tree diagram A tree-shaped diagram depicting the relationship of species, based on comparisons of DNA or protein sequences.

Monoclonal antibody A single antibody type typically produced from an artificial type of fused cell called a hybridoma, which is a cancer cell joined with a B cell. Monoclonal antibodies have many applications.

Monocot A plant with one seed leaf that appears as the plantlet emerges from the soil.

Monohybrid An individual heterozygous for a particular gene.

Monosomy A cell missing one chromosome.

Monozygotic twins Identical twins.

Morula The preembryonic stage consisting of a solid ball of cells.

Motif A DNA sequence common among several transcription factors that folds the factors into characteristic conformations that provide their functions.

Mucopolysaccharides Long carbohydrate molecules that cannot be broken down by an enzyme that is missing in Hurler syndrome.

Mullerian-inhibiting substance A protein activated in the embryo that destroys rudiments of a female reproductive tract, thereby shifting development towards maleness.

Multifactorial A trait molded by several genes and environmental influences.

Multiregional hypothesis The theory that *Homo erectus* gave rise to various geographically widespread populations of humans.

Mutagen A substance that damages or causes change in DNA.

Mutant A phenotype caused by a change in a gene.

Mutant selection A technique of growing cells in the presence of a toxic agent, such as a herbicide, so resistant cells are selected for survival.

Mutation A change in a gene or chromosome.

Myoblast transfer therapy A somatic gene therapy for some types of muscular dystrophy in which immature skeletal (voluntary) muscle cells called myoblasts are transferred to affected muscles where they replace absent or abnormal dystrophin.

Natural selection The tendency for an individual with a certain phenotype in a particular environment to survive and reproduce.

Negative eugenics Interfering with or preventing the reproduction of individuals who are considered to be inferior in some way.

Neonatology The study of the newborn.

Neural tube A tube of tissue that forms down the longitudinal axis of the human embryo during the fourth week of prenatal development. It eventually houses nervous tissue.

Neural tube defect A birth defect in which part of the brain or spinal cord protrudes through the back or neck. Failure of the neural tube to close during the fourth week of prenatal existence causes the neural tube defect.

Neurofibrillary tangles Deposits of the protein tau found in the brains of people with Alzheimer's disease. Tau disrupts microtubules in neuron branches.

Nondisjunction The unequal partition of chromosomes into gametes during meiosis.

Nonheritable (somatic) gene therapy Altering a gene in a somatic cell, so only that tissue in the individual is affected, and the change is not transmitted to the next generation.

Nonreciprocal translocation A piece of one chromosome becomes attached to a non-homologous chromosome.

Notochord A semirigid rod running down the length of an animal's body.

Nucleic acid Deoxyribonucleic acid and ribonucleic acid are nucleic acids. A nucleic acid is an organic molecule consisting of purine and pyrimidine base pairs joined by a sugar-phosphate backbone.

Nuclein A substance in cell nuclei identified in 1871 that was later discovered to be DNA.

Nucleolar organizing region Portions of chromosomes that coalesce to form the nucleolus, where ribosomal subunits are stored.

Nucleolus A structure within the nucleus where RNA nucleotides are stored.

Nucleosome DNA wrapped around an octet of histone proteins.

Nucleotide The building block of a nucleic acid consisting of a phosphate group, a nitrogenous base, and a 5-carbon sugar.

Nucleus A membrane-bound sac in a eukaryotic cell that contains the genetic material.

Oncogene A gene that normally controls cell division but when overexpressed leads to cancer.

Oocyte The female sex cell before it is fertilized.

Oogenesis The differentiation of an egg cell from a diploid oogonium, to a primary oocyte, to two haploid secondary oocytes, to ootids, and finally, after fertilization, to a mature ovum.

Oogonium The diploid cell in the female that gives rise to the egg.

Operon Genes whose enzyme products interact in a coordinated fashion in a particular metabolic pathway.

Opportunistic infection An infection that is rarely seen in individuals with functioning immune systems but appears more often in those with suppressed immunity. AIDS patients suffer many opportunistic infections.

Organelles Specialized structures in eukaryotic cells that carry out specific functions.

Ovaries The paired, female gonads that house developing oocytes.

Ovulation Release of a secondary oocyte from the ovary.

p53 A tumor suppressor gene whose loss of function is implicated in a number of different types of cancers. The p53 gene normally encodes a transcription factor.

Paleontologist A scientist who studies past life.

Paracentric inversion An inverted chromosome not including the centromere.

Parental generation The first generation considered when following transmission of an inherited trait.

Parsimony analysis A statistical method used to identify an evolutionary tree that is likely to represent reality. Such trees are constructed from the degree of similarity of DNA or protein sequences.

Particle bombardment Using a gunlike device to shoot tiny metal particles coated with DNA into cells.

Passive immunity Temporary immunity in a newborn provided by maternal antibodies received before birth or in the mother's milk.

Pedigree A chart showing the relationships of relatives and indicating which ones have a particular trait.

Penetrance The percentage of individuals inheriting a genotype who express the corresponding phenotype.

Pericentric inversion An inverted chromosome including the centromere.

Phenocopy An environmentally-caused trait that appears to be inherited.

Phenotype The observable expression of a genotype in a specific environment.

Phenotypic ratio The ratio of phenotypes resulting from a particular cross.

Philadelphia chromosome A chromosome aberration associated with chronic myeloid leukemia consisting of the tip of chromosome 9 translocated to chromosome 22.

Phosphodiester linkage The covalent bonds forming the sugar-phosphate backbone of DNA.

Photoreactivation enzyme An enzyme that harnesses solar energy to break pyrimidine dimers in DNA.

Photosynthesis The series of biochemical reactions that enable plants to harness the energy in sunlight to manufacture nutrient molecules.

Physical map A chromosome map built of ordered landmarks that are a known physical distance from each other.

Placenta A specialized organ, which develops in certain mammals, connecting the mother to unborn offspring.

Plasma cells Cells that develop from B cells that secrete huge amounts of a single antibody type.

Plasmid A small circle of double-stranded DNA found in some bacteria in addition to their DNA, commonly used as a vector for recombinant DNA.

Pleiotropic A genotype with multiple expressions.

Polar body A small cell generated during female meiosis enabling much cytoplasm to be partitioned into just one of the four meiotic products, the ovum.

Poly A tail A stretch of 100 to 200 adenines found at the 3' end of mRNA before it exits the nucleus.

Polygenic A trait determined by more than one gene.

Polymer A long molecule built of similar subunits.

Polymerase chain reaction (PCR) A method directing DNA replication of a gene of interest in a test tube to rapidly produce many copies of the gene.

Polymorphism A DNA sequence at a certain chromosomal locus that varies between individuals.

Polyploid A cell with one or more extra sets of chromosomes.

Polyps Noncancerous growths that resemble blobs extending from tissue.

Ponderosity A measure of weight taking height into account.

Population A group of interbreeding individuals.

Population bottleneck Narrowing of allele diversity resulting from a disaster that kills many members of a population.

Population genetics The study of gene frequencies in different groups of organisms.

Positional cloning Using genetic and cytogenetic clues to physically locate a gene.

Positive eugenics Providing incentives to reproduce for individuals considered to be superior genetically.

Post-replication repair A contingent of enzymes that detects and replaces mismatched bases in newly replicated DNA.

Preembryo A prenatal human during the first two weeks before tissue layers form.

Pre-implantation genetic diagnosis Application of a DNA probe to one cell removed from an 8-cell preembryo. If the probe's binding indicates that a genetic disease has been inherited, the preembryo from which it came is not chosen to complete development.

Pre-mRNA mRNA prior to removal of introns (non-coding regions).

Primary germ layers The three tissue layers of the embryo. Cells in a particular layer differentiate into specific cell types.

Primary immune response The immune system's response to its first meeting with a nonself antigen.

Primary oocyte The diploid cell in the female that undergoes meiosis I.

Primary spermatocyte The diploid cell in the male that undergoes meiosis I.

Primitive streak A pigmented band along the back of a 3-week embryo that develops into the notochord.

Product rule The chance that two genetic events occur equals the product of the chances of each event occurring.

Promoter A control sequence near the start of a gene.

Pronucleus The genetic package of a gamete.

Prophase The first stage of cell division when chromosomes condense and become visible. During prophase of meiosis I, synapsis and crossing over occur.

Protamines Proteins around which DNA wraps in a sperm head.

Protein A long molecule built of amino acids bonded to each other.

Protooncogene A gene that normally controls the cell cycle that, when overexpressed, functions as an oncogene and causes cancer.

Protoplast A plant cell whose cell wall has been removed. Rendering plant cells into protoplasts makes it easier to manipulate them genetically.

Purine A type of organic molecule with a double ring structure, which includes the nitrogenous bases adenine and guanine.

Pyrimidine A type of organic molecule with a single ring structure, which includes the nitrogenous bases cytosine, thymine, and uracil.

Reading frame The starting DNA base from which a polypeptide sequence is read.

Receptor A molecule in or on a cell membrane that has a pocket that fits another molecule whose binding triggers chemical activity in the cell.

Recessive An allele whose expression is masked by the activity of another allele.

Reciprocal translocation A chromosome aberration in which two nonhomologous chromosomes exchange parts, conserving genetic balance but rearranging genes.

Recombinant A gene combination in an offspring that is different from that of either parent.

Recombinant DNA technology Transferring a gene from a cell of a member of one species to the cell of a member of a different species.

Reduction division Meiosis I when the diploid chromosome number is halved.

Repair The ability of DNA to detect and use enzymes to repair errors in DNA replication.

Replacement hypothesis The theory that Africans replaced the Eurasian descendants of *Homo erectus* by 200,000 years ago.

Replicated chromosome A chromosome consisting of two chromatids. The genetic material is duplicated, but the copies are still held together at the centromere.

Replication fork Locally opened portion of a DNA double helix being replicated.

Repulsion The situation in which a chromosome carries a dominant allele of one gene and a recessive allele of the other, and the homolog has a recessive allele of the first and a dominant allele of the second.

Restriction enzyme A bacterial enzyme that cuts DNA at a specific sequence.

Restriction fragment length polymorphism (RFLP) Differences in restriction enzyme cutting sites between individuals at the same site among the chromosomes.

Retrovirus A virus whose genetic material is RNA. It enters a host cell where the viral enzyme reverse transcriptase produces DNA from the viral RNA, which then integrates into the host's genome where it can direct reproduction of the virus.

Reverse transcriptase An enzyme encoded by a retrovirus that reverse transcribes DNA from RNA.

Rh factor A blood group antigen, determined by 3 genes.

Ribonucleic acid (RNA) A single-stranded nucleic acid built of nucleotides containing a phosphate, ribose, and nitrogenous bases adenine, guanine, cytosine, and uracil.

Ribose A 5-carbon sugar in RNA.

Ribosomal RNA RNA that, along with proteins, comprises the ribosome.

Ribosome A structure built of RNA and protein upon which a gene's message (mRNA) anchors during protein synthesis.

Ribozyme RNA component of an RNA-protein complex that edits introns out of DNA.

Risk factor A characteristic that makes a person more likely to develop a particular illness than someone who does not have that characteristic.

RNA polymerase An enzyme that synthesizes short pieces of RNA to initiate DNA replication. RNA polymerase also adds RNA nucleotides to a growing RNA chain in transcription.

RNA primer A short sequence of RNA that initiates DNA replication.

S phase The synthesis phase of interphase when DNA is replicated and microtubules are produced from tubulin.

Satellites Characteristic blobs on chromosome tips.

Scrapie A viral illness of sheep that resembles kuru, a human illness.

Second messengers Molecules in the cell membrane that react to an incoming message, triggering a cascade of biochemical activity culminating in a cellular response.

Secondary immune response The immune system's response to its second meeting with a nonself antigen.

Secondary oocyte A product of meiosis I in the female.

Secondary spermatocyte A product of meiosis I in the male.

Segregation The distribution of alleles of a gene into separate gametes during meiosis, constituting Mendel's first law.

Self antigens Molecules on cells of the body perceived by the immune system as self and, therefore, not normally provoking an immune attack.

Semiconservative DNA replication The synthesis of new DNA against the separated strands of a parental double helix.

Sense RNA RNA that is translated into protein.

Sense strand The side of the DNA double helix for a particular gene that is transcribed.

Sequence tagged sites Unique (unrepeated) DNA sequences that serve as landmarks on chromosome maps.

Sex chromosome A chromosome that contains the genes determining sex.

Sex-influenced inheritance An allele that is dominant in one sex but recessive in the other.

Sex limited trait A trait affecting a structure or function of the body that is present in only one sex.

Sex linked A gene located on the X chromosome or a trait that results from the activity of such a gene.

Sexual reproduction The combination of genetic material from two individuals to create a third individual.

Signal transduction A cell's receiving and interpreting chemical messages from outside the cell.

Small nuclear ribonucleoproteins The protein portion of the RNA-protein complex that edits introns from genes.

Somatic cell A body cell; a cell other than the sperm or ovum.

Somatic embryo An embryo that develops from a somatic cell. Somatic embryos form from plants.

Somatic mutation A mutation occurring in a somatic (nonsex) cell.

Southern blotting The use of DNA probes to identify specific fragments of DNA.

Species A group of similar individuals that interbreed in nature and are reproductively isolated from all other such groups.

Spectrin A type of cytoskeletal protein that gives support on the inside face of cell membranes.

Sperm The mature male sex cell.

Spermatid Immature sperm cell prior to assuming its final form but after completing meiosis.

Spermatogenesis The differentiation of a sperm cell from a diploid spermatogonium to primary spermatocyte, two haploid secondary spermatocytes, spermatids, and finally to mature spermatozoa.

Spermatogonium A diploid cell in the testes that divides meiotically, yielding daughter cells that become sperm.

Spindle apparatus A structure built of microtubules that aligns and separates chromosomes in cell division.

Spliceosome Complex of small RNAs (ribozymes) and small nuclear ribonucleoproteins that splices introns from genes.

SRY A gene on the Y chromosome that determines maleness by activating certain other genes. SRY stands for sex-determining region of the Y.

Stem cell A cell that divides often.

Submetacentric A chromosome with the centromere in a position that establishes one long arm and one short arm.

Substance P A nervous system messenger biochemical that might be used in preventing or treating Alzheimer's disease by decreasing beta amyloid buildup.

Suppressor T cells Lymphocytes that inhibit the response of all lymphocytes to nonself antigens, shutting off the immune response once the infection has been fought.

Surrogate mother A woman who carries an unborn child for another woman. A gestational surrogate carries a child conceived by another couple. A gestational and genetic surrogate also provides the ovum.

Synapsis The gene-by-gene alignment of homologous chromosomes during prophase of meiosis I.

Synteny The direct comparison of known gene order between species that provides information on evolutionary relatedness.

T cell A type of lymphocyte that produces cytokines in an immune response.

T cell receptor Surface peptides on T cells that bind foreign antigens.

Tau A protein that disrupts the microtubules in neuron processes, and accumulates as neurofibrillary tangles in the brains of people with Alzheimer's disease.

Taxol A biochemical derived from bark and other tissues of the Pacific yew tree that alters cell division by preventing tubulin protein subunits of the spindle apparatus from disassembling. It is being investigated as a cancer treatment.

Telomere The tip of a chromosome.

Telophase The final stage of cell division when two cells form from one and the spindle is disassembled.

Teratogen A chemical or other environmental agent that causes a birth defect.

Tertiary structure The shape assumed by a protein caused by chemical attractions between amino acids that are far apart in the primary structure.

Test cross Crossing an individual of unknown genotype to a homozygous recessive individual.

Testes The paired, male gonads containing the seminiferous tubules, in which sperm are manufactured.

Thymidine kinase An enzyme encoded by a herpes virus. It is used in genetic engineering because a cell that can be altered to manufacture this enzyme can be destroyed by a drug called ganciclovir.

Thymine One of the two pyrimidine bases in DNA.

Tissue In multicellular organisms, groups of cells with related functions.

Tissue typing Identifying the cell surface proteins, or the human leukocyte antigen complex (HLA), encoded by a closely linked set of genes on the short arm of chromosome 6.

Topoisomerases Enzymes that break, untwist, and reattach DNA during replication.

Transcription Manufacturing RNA from DNA.

Transcription factor A protein that controls activation of transcription of other genes.

Transfer RNA A small RNA molecule that binds an amino acid at one site and an mRNA codon at another site.

Transgenic organism Genetic engineering of a gamete or fertilized ovum leading to development of an individual with the alteration in every cell.

Translation Assembly of an amino acid chain according to the sequence of base triplets in a molecule of mRNA.

Translocation Exchange of genetic material between nonhomologous chromosomes.

Translocation carrier An individual with exchanged chromosomes but no symptoms because the total amount of genetic material is not altered.

Transmission genetics The study of the passage of traits between generations.

Transplantable The characteristic of a cancer cell that if transplanted to another individual, a tumor will result.

Triploid A cell with three complete sets of chromosomes.

Trisomy A cell with one extra chromosome.

Trophoblast A layer of cells in the preembryo that develops into the chorion and then the placenta.

Tubulins Hollow protein tubules that assemble into microtubules, which function in cell division.

Tumor necrosis factor An immune system biochemical secreted by T cells that responds to cancer cells and is present in septic shock.

Tumor suppressor A recessive gene whose wild type (normal) function is to hold action of a gene promoting cell division in check.

Ultrasound scanning A procedure that bounces sound waves off of a fetus to form an image on a screen.

Uniparental disomy Inheriting both copies of a gene or chromosome region from one parent.

Unreplicated chromosome A chromosome consisting of one chromatid (one DNA double helix).

Uracil One of the two pyrimidine bases in RNA.

Uterus The muscular, saclike organ in the human female in which the embryo and fetus develop.

Vaccine A portion of an infectious organism or agent that is sufficient to stimulate the immune system of the host to mount an attack without producing symptoms.

Vaccinia The virus that causes smallpox.

Variable number of tandem repeats (VNTRs) A DNA sequence at a certain chromosomal locus that repeats a different number of times in different individuals.

Variably expressive A phenotype that varies in intensity in different individuals, perhaps reflecting the influences of other genes.

Vas deferens In the human male, a tube from the epididymis that continues to become the vas deferens, which joins the urethra in the penis.

Vector A piece of DNA to which DNA from one type of organism can be attached and transferred into a cell of another organism.

Viroid Infectious genetic material.

Virus An infectious particle consisting of a nucleic acid (DNA or RNA) wrapped in protein.

Whole chromosome paints DNA probes that cover an entire chromosome, enabling it to be highlighted in a chromosome preparation.

Wild type A phenotype or allele that is the most common for a certain gene in a population: a "normal" phenotype.

X inactivation The turning off of one X chromosome in each cell of a female mammal at a certain point in prenatal development.

Yeast artificial chromosome (YAC) Pieces of yeast chromosomes into which foreign DNA is inserted.

Yolk sac An extraembryonic membrane that manufactures blood cells.

Zinc finger A DNA-binding region of many transcription factors consisting of amino acids looped around zinc atoms.

Zona drilling Gently piercing the outer layer of an oocyte so that sperm can more easily enter in vitro.

Zona pellucida A thin, clear layer of proteins and sugars surrounding a secondary oocyte.

Zygote In prenatal humans, the organism during the first two weeks of development. Also called a preembryo.

Zygote intrafallopian transfer (ZIFT) A technology in which a preembryo fertilized in vitro is introduced into the woman's Fallopian tube, rather than into her uterus.

A GLOSSARY OF CHEMICAL TERMS RELEVANT TO GENETICS

Acid A molecule that releases hydrogen ions (OH⁻) into water. DNA and RNA are acids.

Amino acid An organic molecule built of a central carbon atom bonded to a hydrogen atom, an amino group, a carboxylic acid, and a group, designated "R," that can vary. A polymer of amino acids is a peptide or a polypeptide. A protein consists of one or more polypeptides.

Atom A chemical unit composed of protons, neutrons, and electrons that cannot be further broken down by chemical means.

Base A molecule that releases hydroxide ions (OH⁻) into water.

Carbohydrate A compound containing carbon, hydrogen, and oxygen with twice as many hydrogens as oxygens. Carbohydrates include sugars and starches.

Compound A molecule consisting of different atoms.

Covalent bond An attractive force between atoms formed by the sharing of electrons between them.

Dehydration synthesis Formation of a covalent bond between two molecules by the loss of water.

Disaccharide A sugar built of two bonded monosaccharides. Disaccharides include sucrose, maltose, and lactose.

Electron A subatomic particle carrying a negative electrical charge and a negligible mass, which orbits the atomic nucleus.

Electron-transport chain A series of linked chemical reactions in which an electron is lost by one compound and picked up by another. These are oxidation-reduction reactions.

Element A pure substance consisting of atoms containing a characteristic number of protons.

Glycolysis A catabolic (breakdown) pathway occurring in the cytoplasm of all cells. In glycolysis one molecule of glucose is split and rearranged into two molecules of pyruvic acid.

Hydrocarbon A molecule consisting of only carbon and hydrogen.

Hydrogen bond A weak chemical bond between negatively charged portions of molecules and hydrogen ions.

Hydrolysis reaction The splitting of a molecule in two by adding water.

Hydrophilic Attraction of part of a molecule to water.

Hydrophobic Repulsion of part of a molecule from water.

Ion An atom that has lost or gained electrons giving it an electrical charge.

Ionic bond Attraction between oppositely-charged ions. Atoms combine to complete their outermost electron shells.

Isotope A differently weighted form of an element. A biochemical incorporating an unusual isotope can be used to trace biological activities.

Lipid An organic molecule that is insoluble in water. Fats and oils are lipids.

Macromolecule A large chemical structure built of atoms bonded to one another. DNA, RNA, and proteins are macromolecules.

Molecule A structure resulting from the combination of atoms.

Monosaccharide A sugar built of one 5- or 6-carbon unit, including the dietary sugars glucose—galactose and fructose—and the sugars that are part of nucleic acids—ribose (in RNA) and deoxyribose (in DNA).

Neutron A particle in an atom's nucleus that is electrically neutral and has one mass unit.

Nucleus (atomic) The central region of an atom, consisting of protons and neutrons.

Oxidation reaction A chemical reaction in which electrons are lost.

Peptide bond A chemical bond between two amino acids resulting from dehydration synthesis.

pH scale A measurement of how acidic or basic a solution is. pH is the hydrogen ion concentration.

Phospholipid A molecule built of a lipid and a phosphate that is hydrophobic at one end and hydrophilic at the other end. Phospholipids are integral components of cell membranes.

Proton A particle in an atom's nucleus carrying a positive charge and having one mass unit.

Reduction reaction A chemical reaction in which electrons are gained.

Simple carbohydrates Monosaccharides and disaccharides (sugars).

Solution A homogenous mixture of a substance (the solute) dissolved in water (the solvent).

A GLOSSARY OF DISORDERS MENTIONED IN THE TEXT

designations: ar = autosomal recessive
AD = autosomal dominant
xlr = sex-linked recessive
XLD = sex-linked dominant
mf = multifactorial
chromosome abnormalities

Achondroplasia (AD) A form of dwarfism in which the limbs are stunted, but the head and trunk are normal size.

Acquired immune deficiency syndrome A progressive loss of immune function believed to be caused by infection with the human immunodeficiency virus.

Acute promyelocytic leukemia A type of white blood cell cancer associated with a fusion protein that is translated when a translocation between chromosomes 15 and 17 brings together a gene encoding the retinoic acid receptor and an oncogene.

Adrenoleukodystrophy (ar, xlr) In the autosomal recessive form, neurological impairment leads to death by age 3. The adrenal glands degenerate, and there are characteristic facial features. The sex-linked form is less severe.

African dietary iron overload (mf) The tendency to excessively store iron in the blood, caused by heredity and the environment.

Agammaglobulinemia (xlr) A lack of a class of antibodies impairs immunity.

Alpha-1-antitrypsin deficiency (AD) Hereditary emphysema caused by deficient alpha-1-antitrypsin, which normally keeps levels of elastase in check. Elevated elastase destroys lung tissue.

Alport syndrome (ar, AD, xlr) Hearing loss and inflamed kidneys due to a defect in collagen, a connective tissue protein.

Alzheimer disease (AD, mf) A progressive loss of memory and reasoning ability caused by buildup of gummy proteins in the brain, beginning in mid to late adulthood.

Amelogenesis imperfecta (xlr) Tooth enamel is soft and white.

Amyloidosis (Dutch variety) (xlr, ar, AD) A gummy protein called amyloid cannot be normally processed. It builds up in the brain causing cerebral hemorrhage between ages 45 and 65.

Angelman syndrome (chromosome 15 deletion) A rare syndrome of mental retardation with a puppet-like appearance (poor muscle tone, large protruding tongue, extreme and inappropriate laughter, convulsions). Both chromosome 15s have a small deletion and are usually both of paternal origin, exhibiting genomic imprinting.

Aniridia (AD) Absence of the iris.

Ankylosing spondylitis (ar) An HLA-linked disorder in which the vertebrae are inflamed and deformed.

Aortic aneurysm (AD) The aorta balloons out and bursts, caused by mutation in a collagen (connective tissue protein) gene.

Ataxia telangiectasis (ar) Deficiency of topoisomerases prevents DNA strands from untwisting during replication. Symptoms include lack of muscular coordination, characteristic skin spots from dilated capillaries, high risk of cancer, involuntary eye movements, and frequent sinus and lung infections.

Becker muscular dystrophy (xlr) A late-onset form of muscular dystrophy caused by a mutation in the gene that encodes dystrophin, a protein that ties together bundles of other muscle proteins.

Beckwith-Wiedemann syndrome A syndrome of kidney and adrenal tumors, enlarged tongue, and other symptoms caused by inheriting two copies of one parent's genes in certain somatic cells.

Biotinidase deficiency (ar) Inability to metabolize the vitamin biotin leads to mental retardation; seizures; a rash; loss of vision, hearing, and hair; and slow growth and progresses to convulsions, coma, and death if not treated.

Bloom syndrome (ar) Inactive or heat sensitive DNA ligase slows DNA replication. An affected child is very small at birth and has characteristic facial features, a skin rash on exposure to sun, high risk of cancer, and greatly impaired immunity.

Burkitt's lymphoma A cancer common in Africa in which a large tumor develops from lymph glands near the jaw. Burkitt's lymphoma is often triggered by the Epstein-Barr virus, which causes a translocation placing an oncogene on chromosome 8 next to an antibody gene on chromosome 14. When the antibody gene is highly expressed, so is the oncogene. Cancer begins.

Cat eye syndrome (ring chromosome 22 resulting in partial trisomy) Vertical pupils, mental retardation, urinary tract anomalies, and skin over the anus.

Charcot-Marie-Tooth disease (xlr) Loss of feeling in the ends of the extremities. Effects vary greatly from person to person.

Chondrodysplasia (ar) Stunted growth and deformed joints due to mutation in a gene for collagen, a connective tissue protein.

Chronic granulomatous disease (xlr) A disorder of white blood cells leading to frequent skin and lung infection and enlarged liver and spleen.

Chronic myeloid leukemia A type of white blood cell cancer, nearly always associated with the Philadelphia chromosome, consisting of the tip of chromosome 9 translocated to chromosome 22.

Cleft palate (xlr, ar, AD, mf) Failure of the bones forming the roof of the mouth to close.

Common acute lymphoblastic leukemia (mf) A white blood cell cancer that is more prevalent among children who have not had many infections and whose unchallenged immune systems may, therefore, be less capable of fighting cancer cells.

Congenital adrenal hyperplasia (ar) A hormonal abnormality in which female genitalia are masculinized and males go through precocious puberty. Growth is accelerated although ultimate height is below normal. Sex is often difficult to tell.

Congenital hypothyroidism (ar) Deficiency of thyroid hormone causes mental retardation, poor growth, hearing loss, and other neurological impairment.

Cri-du-chat syndrome (deletion 5p) An affected child has a shrill cry like a cat's, mental retardation, widely spaced eyes, rounded face, and small head.

Cystic fibrosis (ar) In many tissues, abnormal chloride channels cannot migrate from a cell's interior to the cell membrane where they would normally monitor the travel of sodium and chloride ions. Salt trapped inside cells lining certain glands, such as in the lungs, pancreas, skin, and small intestine, causes water to flow in, drying out mucous secretions. The blockage of these organs by sticky mucus causes the symptoms of frequent respiratory infections, difficulty breathing, poor weight gain, and salty sweat.

Diabetes insipidus (ar, AD) Copious urination caused by suppression of antidiuretic hormone.

Distal symphalangism (AD) Stiffened fingers and toes with tiny nails.

Down syndrome (trisomy 21 or translocation 21) A syndrome of mental retardation, characteristic facial features, palm creases, poor immunity, high susceptibility to leukemia, deafness, and other health problems, caused by extra material from chromosome 21.

Duchenne muscular dystrophy (xlr) Progressive muscle weakness caused by absent or abnormal dystrophin, a protein that holds together muscle bundles.

Dystrophic epidermolysis bullosa (ar) Due to a breakdown in the collagen fibrils that attach the outer to the inner layer of skin, the skin blisters when touched.

Edward syndrome (trisomy 18) An extra chromosome 18 causes mental and physical retardation, skull and facial abnormalities, extreme muscle tone that twists the head oddly, and defects in all organ systems.

Ehlers-Danlos syndrome (ar, AD, xlr) Joints are lax and skin stretches greatly, very easily. Caused by mutations in genes for collagen, a connective tissue protein.

Factor XI deficiency (ar) A mutation in the gene encoding the protein portion of a plasma glycoprotein interferes with the early steps of blood clotting and leads to severe bleeding with injury or surgery.

Familial hypercholesterolemia (AD) Cells in the liver lack low density lipoprotein receptors, causing buildup of cholesterol on artery walls, resulting in early, severe heart disease.

Familial polyposis of the colon (AD) Many polyps and tumors, benign and sometimes cancerous, develop along the intestinal lining. (Also known as familial adenomatus polyposis.)

Fanconi anemia (ar) A point mutation leads to chromosome breaks. Deficient excision repair causes severe depletion of many blood cell precursors in the bone marrow. Risk of leukemia is very high. Other symptoms include skin color changes and malformations of the limbs, kidneys, and heart.

Favism (G6PD deficiency) (xlr) Anemia that occurs when a person with too little G6PD enzyme eats fava beans or takes certain drugs.

Fragile X syndrome A form of mental retardation accompanied by large testicles and a characteristic long face, associated with fragile sites on the X chromosome. It may occur in a milder form in, or be passed by, females.

Galactosemia (ar) The inability to metabolize the sugar galactose leads to muscle weakness, cerebral palsy, seizures, mental retardation, cataracts, and liver disease.

Gardner syndrome (AD) A syndrome including colon polyps, cancer, extra teeth, and pigment patches in the eye.

Gaucher disease (ar) A lysosomal storage disease causing deficiency of glucocerebrosidase. This results in swelling of the liver and spleen, easy falling, anemia, internal bleeding, and early death.

Glioma A deadly type of brain tumor affecting the supportive glial cells that surround nerve cells. Glioma is being treated experimentally with gene therapy.

Gout (xlr, AD, mf) Inability to metabolize uric acid precursors leads to sandy deposits of uric acid in the joints and kidneys.

Green colorblindness (xlr) Abnormal green cone pigments in the retina impair vision of these wavelengths.

Greig syndrome (AD) A syndrome of extra and fused toes and fingers, broad nose and thumbs, large head, and mild developmental delay. Can be associated with a small deletion in chromosome 7.

Hemochromatosis (ar) A defect in iron metabolism in which the body retains excess iron. Symptoms include diabetes, proneness to infection, cirrhosis of the liver, excess skin pigmentation, and heart failure.

Hemophilia A (xlr) Deficiency of clotting factor VIII causes internal bleeding and slowed clotting in wounds.

Hemophilia B (xlr) Deficiency of clotting factor IX causes mild impairment of blood clotting.

Hereditary spherocytosis (ar) Ankyrin proteins are abnormal in the red blood cell membrane, causing it to disintegrate, resulting in anemia and spleen damage.

Hirschsprung disease (ar) Missing nerves in the rectum and colon lead to severe constipation. Other symptoms include mental retardation and autism.

Homocystinuria (ar) Inability to metabolize the amino acid methionine causes blood clots, thin bones, mental retardation, seizures, muscle weakness, and mental disturbances.

Hunter syndrome (xlr) A syndrome of mental retardation and other brain damage that is lethal before adulthood. It is also known as mucopolysaccharidosis II and can be caused by a small deletion.

Huntington disease (AD) A degenerative neurological disorder, usually beginning in the late thirties, including personality changes and uncontrollable movements.

Hurler disease (ar) A lysosomal storage disease in which deficiency of alpha-L-iduronidase causes bone deformities, mental retardation, and other symptoms.

Hydrocephalus (ar, AD, xlr, mf) Swelling of the ventricles in the brain with cerebrospinal fluid before birth may lead to mental retardation if the pressure is not removed shortly after birth.

Hydrops fetalis An attack of a second or subsequent Rh+ fetus by the immune system of a mother who is Rh–.

Hypertension (mf) Elevated blood pressure caused by the interactions of several genes and environmental factors.

Hypohydrotic ectodermal dysplasia (xlr) Absence of teeth, hair, and sweat glands.

Hypophosphatemia (XLD) A malfunctioning transcription factor prevents bone cells from responding to vitamin D, causing the weakened bones of vitamin D resistant rickets.

Ichthyosis (xlr) Rough, scaly skin on the scalp, ears, neck, abdomen, and legs.

Incontinentia pigmenti (XLD) Affected females have a marbled skin coloration and are born with vesicles that turn wart-like. Hair falls out, teeth are abnormal, and there are seizures. Affected males usually die in utero.

Jacob syndrome (XYY) Once thought to be associated with criminality and aggressive tendencies, males with an extra Y chromosome are now believed to only have great height and acne.

Juvenile diabetes (mf) A deficiency of the pancreatic hormone insulin impairs transport of glucose in the blood to cells that would use it. High blood sugar causes symptoms of excessive thirst and urination and weight loss.

Kallmann syndrome (xlr) A syndrome consisting of the inability to smell and underdeveloped testes.

Kaposi's sarcoma A type of soft tissue cancer seen often in AIDS patients. It produces characteristic purplish skin lesions.

Klinefelter syndrome (XXY) An extra X chromosome in a male results in lack of development of secondary sexual characteristics, infertility, long limbs, great height, sometimes breast development, and sometimes impaired mental development.

Lactose intolerance (AD) The inability to digest lactose, leading to cramps after eating.

Lesch-Nyhan syndrome (xlr) Absence of the enzyme HGPRT causes mental retardation, uric acid crystals in the urine, other physical symptoms, and behavioral symptoms such as self-mutilation. The body cannot recycle two of the four types of DNA nitrogenous bases, leading to buildup of uric acid.

Li-Fraumeni family cancer syndrome (AD) A condition in which a mutant p53 gene is inherited in the germline, and somatic mutations cause cancer in a variety of tissues.

Lissencephaly (ar) The surface of the cerebrum, normally convoluted, is smooth. Symptoms include profound mental retardation, seizures, and other neurological problems.

Malignant hyperthermia (AD) High fever, muscle contraction, swelling, and pressure on the brain, followed by coma and death. Exposure to a certain anesthetic drug is necessary for malignant hyperthermia to be expressed.

Maple syrup urine disease (ar) The inability to metabolize the amino acids leucine, isoleucine, and valine. Symptoms include lethargy, vomiting, irritability, mental retardation, coma, and death by one month if not treated. The urine has a characteristic sweet odor.

Marfan syndrome (AD) Mutant or absent fibrillin, an elastic connective tissue protein, causes varied symptoms of a weakened aorta wall, long limbs, caved in chest, lens dislocation, and spindly fingers.

Melanoma (mf) A deadly type of skin cancer that begins with an irregularly shaped, pigmented patch of skin, and spreads quickly to the bloodstream. At least one gene contributes to its development.

Menkes disease (xlr) Abnormal transport of copper leads to kinky hair and brain atrophy.

Myotonic dystrophy (AD) A disorder of muscle wasting that grows in severity with each generation, as the gene enlarges.

Nail-patella syndrome (AD) Odd-shaped nails and kneecaps.

Neurofibromatosis type I (AD) Brown pigment marks (cafe au lait spots) appear on the skin, and numerous tumors grow beneath the skin. Tumors are usually benign but may be malignant. It is caused by a disruption in the control of signal transduction of a growth factor signal into the cell, causing uncontrolled cell division.

Neurological neoplastic syndrome An autoimmune condition that sometimes arises in cancer patients in whom cancer cell surfaces resemble body cell surfaces, triggering immune attack against healthy body cells.

Ornithine transcarbamylase deficiency (xlr) Mental deterioration caused by accumulation of ammonia in the blood.

Osteoarthritis (AD) Joint inflammation and destruction caused by mutation in genes for collagen, a connective tissue protein.

Osteogenesis imperfecta (ar, AD) Absent or abnormal collagen leads to very fragile bones, hearing loss, and blue eye sclera.

Osteoporosis (ar, mf) Bones weaken due to mutation in collagen genes.

Patau syndrome (trisomy 13) An extra chromosome 13 causes a cleft lip, a large triangular nose, extra fingers and toes, skull and facial abnormalities, physical and mental retardation, and other defects in all organ systems.

Phenylketonuria (ar) An inborn error of metabolism in which a deficit of phenylalanine hydroxylase causes a buildup of phenylalanine that results in mental retardation, fair skin, and other symptoms.

Phocomelia (ar) Failure of the limbs to develop completely. Limb birth defects caused by the drug thalidomide are a phenocopy of phocomelia.

Pneumocystis carinii **pneumonia** A lung infection commonly seen in AIDS patients caused by a protozoan.

Polycystic kidney disease (AD) Cysts develop in the kidneys in early adulthood, followed by bloody urine, high blood pressure, and abdominal pain.

Polydactyly (AD) Extra fingers and/or toes.

Pompe disease (ar) A lysosomal storage disease in which glycogen accumulates in muscle and liver cells, causing heart failure.

Porphyria (ar) An inborn error of metabolism in which a missing enzyme prevents metabolism of red blood cell breakdown products, leading to a wide variety of symptoms that occur in a specific sequence, including: abdominal pain, red urine, fever, headache, hoarseness, and eventually coma and death.

Porphyria variegata (AD) An enzyme anomaly causes a severe reaction to barbiturate anesthetics.

Prader-Willi syndrome (deletion chromosome 15) A rare syndrome of mental retardation, small hands and feet, and no sexual maturity. Prader-Willi syndrome is associated with a homozygous deletion in both chromosome 15s, which are both of maternal origin. Therefore, this condition exhibits genomic imprinting.

Primary adrenal hypoplasia (ar, xlr) Great disorganization of the adrenal glands, causing hormone deficiencies with wide-ranging effects. Death occurs in early childhood without hormone therapy. Puberty is delayed or absent, and mental retardation is possible.

Progeria (ar) An inherited, accelerated aging disease.

Retinitis pigmentosa (xlr) A degeneration of the retina, causing night blindness, constriction of the visual field, and clumps of pigment in the eye.

Retinoblastoma (AD) A rare childhood eye cancer caused by loss of function of a tumor suppressor gene. Retinoblastoma occurs in sporadic and inherited forms depending upon whether there are two somatic mutations or one somatic and one germinal mutation, respectively.

Schizophrenia (mf) Loss of the ability to organize and interpret thoughts and perceptions, leading to withdrawal from reality and inappropriate actions and behaviors.

Severe combined immune deficiency (ar) A class of inborn errors of metabolism in which a deficiency of adenosine deaminase leads to a lack of T and B cells, severely impairing immune function.

Sickle cell disease (ar) A point mutation in the beta globin gene causes hemoglobin to change its conformation under conditions of low oxygen. This bends the red blood cell into a sickle shape. As the sickle-shaped cells become caught in blood vessels, circulation is blocked, leading to anemia, high risk of recurrent infections, joint pain, a swollen, overworked spleen, and other symptoms.

Stickler syndrome (AD) Joint pain and degeneration of the fluid surrounding the retina and in the retina, caused by a mutation in a collagen gene.

Tay-Sachs disease (ar) A mutation in the gene encoding hexoseaminidase causes profound nervous system degeneration that kills in early childhood. The child is born with a characteristic cherry red spot in the retina, but behaves normally for the first few months. Gradually, vision, hearing, and mobility are lost.

Testicular feminization (xlr) A male embryo does not respond to male hormones and continues developing female reproductive organs even though the chromosome constitution is male (XY).

Triplo-X (XXX) A female with an extra X chromosome may have irregular menstrual periods, be very tall, and is at increased risk of producing gametes that have unbalanced sex chromosomes.

Turner syndrome (XO) A female with one X chromosome. Many Turner embryos and fetuses die in utero, but those born have few symptoms—short stature, widely spaced nipples, webbed skin on the back of the neck, and lack of development of secondary sexual characteristics. A Turner female is infertile.

Type I hyperlipoproteinemia (ar) Deficiency of lipoprotein lipase, causing an increase in triglycerides in the blood.

Von Willebrand disease (ar, AD, xlr) Mutation in the gene for the von Willebrand clotting factor causes a person to bruise easily because capillaries break easily.

Waardenburg syndrome (AD) A syndrome characterized by a white forelock, wide-spaced light-colored eyes, and hearing impairment.

Wilms tumor (AD) A childhood cancer of the kidneys caused by absence of a tumor suppressor gene.

Wilson disease (ar) Inability to metabolize copper leads to stomachache, headache, liver inflammation, loss of balance, a gravelly voice, changed handwriting, an uncontrollable facial expression, drooling, and eventually death if not treated with penicillamine to rid the body of excess copper.

Wiskott-Aldrich syndrome (xlr) Bloody diarrhea, infections, rash, too few platelets, and death by age 10.

Xeroderma pigmentosum (ar, AD) A disorder of DNA repair in which sun exposure easily causes mutations and skin cancer.